Casting Processes

Edited by T. R. Vijayaram

Published in London, United Kingdom

Casting Processes
http://dx.doi.org/10.5772/intechopen.97894
Edited by T. R. Vijayaram

Contributors
Narducci Junior Carlos Jr., Yuichi Tsukaguchi, Kodai Fujita, Hideki Murakami, Roderick I.L. Guthrie, Mukkollu Sambasiva Rao, Amitesh Kumar, Vincent Haanappel, Jerzy Zych, Marcin Myszka, Janusz Postuła, A. Babafemi Abiodun Ogunkola, Adeolu Adesoji Adediran, Olanrewaju Seun Adesina, Francis Odikpo Odikpo Edoziuno, Christian Okechukwu, M. Saravana Kumar

First published in London, United Kingdom, 2022 by IntechOpen
IntechOpen is the global imprint of INTECHOPEN LIMITED, registered in England and Wales, registration number: 11086078, 5 Princes Gate Court, London, SW7 2QJ, United Kingdom

British Library Cataloguing-in-Publication Data
A catalogue record for this book is available from the British Library

Additional hard and PDF copies can be obtained from orders@intechopen.com

Casting Processes
Edited by T. R. Vijayaram
p. cm.
Print ISBN 978-1-80355-495-2
Online ISBN 978-1-80355-496-9
eBook (PDF) ISBN 978-1-80355-497-6

Meet the editor

Dr. T. R. Vijayaram is a senior professor in the Department of Mechanical Engineering, School of Mechanical Engineering, Bharath Institute of Higher Education and Research (BIHER), India. He is an expert in the field of materials and metallurgical, manufacturing, and mechanical engineering. He received his BE in Mechanical Engineering from Madurai Kamaraj University, India, followed by an ME in Industrial Metallurgy from PSG College of Technology, Bharathiyar University, India. He obtained his Ph.D. in Mechanical Engineering from Universiti Putra Malaysia. Later, he worked as a rector researcher in metallurgy at the Department of Chemistry and Industrial Chemistry (DCCI), Genoa University, Italy. His passion for academics, research, and education led him to obtain an MBA in Educational Management and an MA in Sociology from the University of Madras, India. He is also a Chartered Engineer (India) and a member of several professional and scientific bodies in India and abroad, including the Indian Society for Technical Education (ISTE), Institution of Engineers (IEI), Institute of Indian Foundryman (IIF), and Society of Automotive Engineers (SAE). He is also a fellow of IEI. Dr. Vijayaram received the Distinguished Scientist Award in Metallurgical and Materials Engineering for his outstanding contribution to metallurgy. He has published more than 210 papers in international and national journals, conferences, broadsheets, and magazines. His areas of research include materials engineering, metallurgical engineering, manufacturing engineering, and mechanical engineering.

Contents

Preface

Metal casting is a basic manufacturing process in which molten metal at a particular temperature is poured into a mold cavity of a particular shape and allowed to solidify completely. The solidified part is often called the casting, which is taken out by breaking the mold. Casting has the best design flexibility when compared to other manufacturing techniques such as rolling, forging, welding, stamping, and extrusion.

The metal casting process also offers advantages in the form of enabling the production and formation of attractive and aesthetic designs, as there is no restriction to the assembly of shapes. Castings or cast components are used in several industries like agriculture, defense, automotive, culinary, mining, construction, health care, and paper. Metals like aluminum, magnesium, copper alloys, tin, lead alloys, iron, and steel are cast in metallic and sand molds. The quality of cast metals and alloys is determined by casting properties like fluidity, shrinkage, porosity, surface finish, and resistance to crack formation.

Different types of casting processes have been developed to manufacture defect-free castings. Foundry technologists have consistently researched several casting techniques to further improve existing processes. Green sand molding mixtures have been tested to optimize the required sand properties to obtain castings of good quality. The molten metal flow is governed by the sand properties. Graphite morphology in cast iron is governed by the solidification rate and inoculation efficiency. Different casting processes produce different casting properties and microstructures in castings. This book discusses various metal casting processes used in the foundry industries. It discusses gravity die casting, semisolid slope casting technology, opportunities and recent trends of the squeeze casting process, novel modeling for molten metal flow in the mold cavity, green sand molding process, and technology of producing cast iron.

Casting Processes includes six chapters in four sections. Section 1 discusses the semi-solid casting and squeezes casting processes. Section 2 depicts the flowability of sand mixtures and liquid metal flow in mold cavities. Section 3 describes the gravity die casting process. Section 4 explains cast iron technology, including information on the morphology of graphite formation, properties, and application. Chapters in these sections discuss the different types of casting processes and their applications in detail.

I would like to thank IntechOpen for giving me the opportunity to serve as editor of this book.

Dr. T.R. Vijayaram
Senior Professor,
Department of Mechanical Engineering,
School of Mechanical Engineering,
BIST,
BIHER, Bharath Institute of Higher Education and Research,
Selaiyur, Chennai, Tamil Nadu, India

Section 1

Semi Solid Casting

Chapter 1

Slope Casting Process: A Review

Mukkollu Sambasiva Rao and Amitesh Kumar

Abstract

Semi solid processing is a near net shape casting process and one of the promising techniques to obtain dendritic free structure of metals. Semi solid casting gives numerous advantages than solid processing and liquid processing. Semi solid casting process gives, Laminar flow filling of die without turbulence, Lower metal temperature, Less shrinkage, Less porosity, Higher mechanical properties. Semi solid casting process is industrially successful, producing a variety of products with good quality. Slope Casting process is a simple technique to produce semi solid feed-stoke with globular microstructure and dendrite free structure castings. Slope casting process depends on different process parameters like slope length, slope angle, pouring temperature etc. The present study mainly focuses on review of various explorations made by researchers with different process parameters of the Slope casting process and explain the mechanisms that lead to microstructural changes which leads to good mechanical properties.

Keywords: semi-solid process, thixo casting, Rheo casting, slope casting, aluminum alloys, non dendritic structure, slope length, slope angle, slope plate temperature, slope vibration

1. Introduction

Semi-solid processing as the name suggests is the processing of non-dendritic material between its liquidus and solidus temperatures. In recent years, much work has been conducted in exploring this field with respect to understanding the mechanisms involved. The inherent properties of semi-solid materials at the semi-solid processing temperature such as lower heat content, relatively higher viscosity comparable to liquids and low flow stresses, enables the semi-solid process to show distinct advantages over fully liquid and/or fully solid-state processes. Some of the important benefits of this technique are low mold erosion, low energy consumption, improved die filling, less gas entrapment, lower solidification shrinkage, reduced macro-segregation and fine microstructure. Therefore, this process is rapidly gaining commercial importance [1–3]; A non-dendritic microstructure can be obtained by stirring, either mechanically or electro-magnetically; grain refining; low superheat melt processing; solid state mechanical treatment and reheating [4–7]; The manufacturing industries widely focused on the semi solid routes to produce components with superior mechanical and metallurgical properties. Slope casting process is one of the simplest techniques to produce semi solid slurry [8]. Slope cating process is pouring of molten metal through a slope channel into a mold. This slope channel help as a site for nucleation and fragmentation of dendrites due to shearing force between different layers of flowing stream [9]. Slope casting process depends on different process parameters like slope length, slope angle, pouring temperature etc. [10–14]. In recent years Aluminum alloys are using mostly in the automotive industries. Among the Aluminum alloys, the Al-Si alloys have good casting characteristics like high fluidity and good cast-ability which makes them advantageous for both small and complicated castings. Every year lakhs of Aluminum alloy components are produced through semi solid processing route. The present study mainly focuses on review of various explorations made by researchers with different process parameters of the Slope casting process and explain the mechanisms that lead to microstructural changes which leads to good mechanical properties.

2. Semi-solid casting

The processing of alloy between liquidus and solidus (mushy zone) range is known as the semi solid process, it was first discovered in 1970s, by spancer at Massachusetts Institute of Technology (MIT); found that at semi solid range of alloy behaves thixotropically (Decreases in viscosity if it is sheared but it will thicken again if it is allowed to stand)and by applying continuous stirring on the semi solid state produced no dendritic and spheroidal microstructure [15–19]. The semi solid casting route gives enormous advantages like dendritic free structure leads to globular structure as seen in **Figure 1**, less defects such as porosity, shrinkage, gas entrapment and macro-segregation. Better advantages than conventional casting that superior quality, low forming temperature, superior mechanical properties with microstructural refinement. The semi-solid process results in a non-dendritic microstructure due to forming at a temperature between solidus and liquidus temperature as shown in phase diagram, **Figure 2**. In semi solid process, temperature has a pivotal role on the resultant microstructure like orientation of grain, morphology of grain during solidification of alloys [22–24]. Semi-solid processing is used for all the shape forming processes which take advantage of the semi-solid range of the alloys for processing. Rheology and Thixotropy, two basic phenomena play a major role, In semi-solid processing. The apparent viscosity of a material in the liquid state varies with change in shear rate In Rheology. This gives the liquid like slurry to be processed even at sufficiently high solid fraction [25]. Thixotropy, is the ability of a material to Decreases

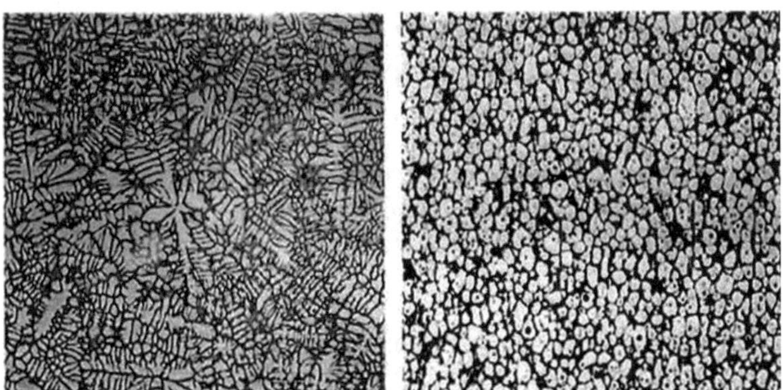

Figure 1.
Using semi solid process technique dendritic structure changes to globular [20].

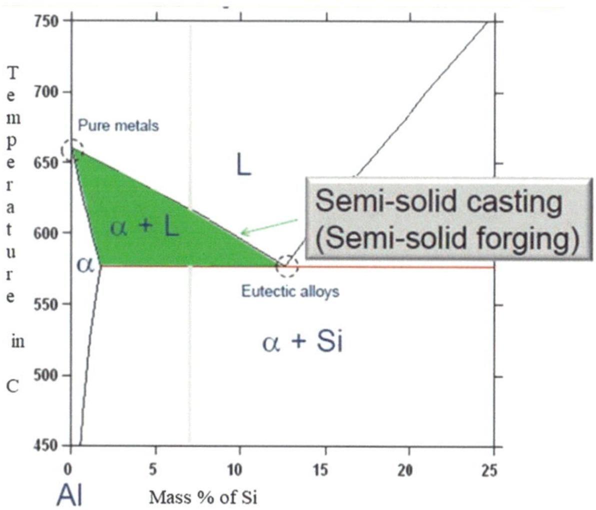

Figure 2.
Phase diagram of Al-Si alloy [21].

in viscosity if it is sheared but it will thicken again if it is allowed to stand [26]; A material with a non-dendritic structure is the best suitable material for semi-solid processing. it is believed that, in the semi-solid state, the non-dendritic equiaxed grains easily slide/glide on each other on the application of a shear force [27].

Thixotropy behavior can be define as when the material state is partially solid with 40–50% solid fraction and is sheared applied by external force, then its viscosity will decrease due to the break/detachment of the coalescence material, and it will flow like a liquid, for a certain time if it is allowed to stand, equiaxed coalescence will increase the viscosity of the material, by that it being able to support its own weight in the same way as if it was solid [28].

2.1 The mechanism of non-dendritic structure

To describe mechanism for non-dendritic structure in semi solid process many theories have been proposed. These mechanisms include dendrite arm fragmentation, dendrite arm root re-melting, and growth control mechanism. Hv Atikson et al. [20], Vogel et al. [29]; proposed that under shearing forces dendrite arms bends due to its plasticity, which introduce large misorientations inside the dendrite arms and dislocations introduced; rearrangements of dislocations occur to form grain boundaries at the melting temperature. The energy of the grain boundaries becomes more than twice the liquid/solid interfacial energy when the misorientations between grain boundaries are more than 20°, then separation of the dendritic arms observes due to wetting of the grain boundaries by liquid metal. Schematically illustrated in **Figure 3**.

Hellawell et al. [30]; proposed grain multiplication theory, Thermal convention and shearing force have a direct effect at the roots of secondary dendrite arms, melting off rather than breaking off secondary arms observed, and grain multiplication, schematically illustrated in **Figure 4**. Evolution of structure during solidification with shear force depends on the cooling rate and shear rate, with increase in shear and cooling rate gives non dendritic/globular structure that that the particle shape and size vary irreversibly with shear and colling rate. Illustrate in **Figure 5**.

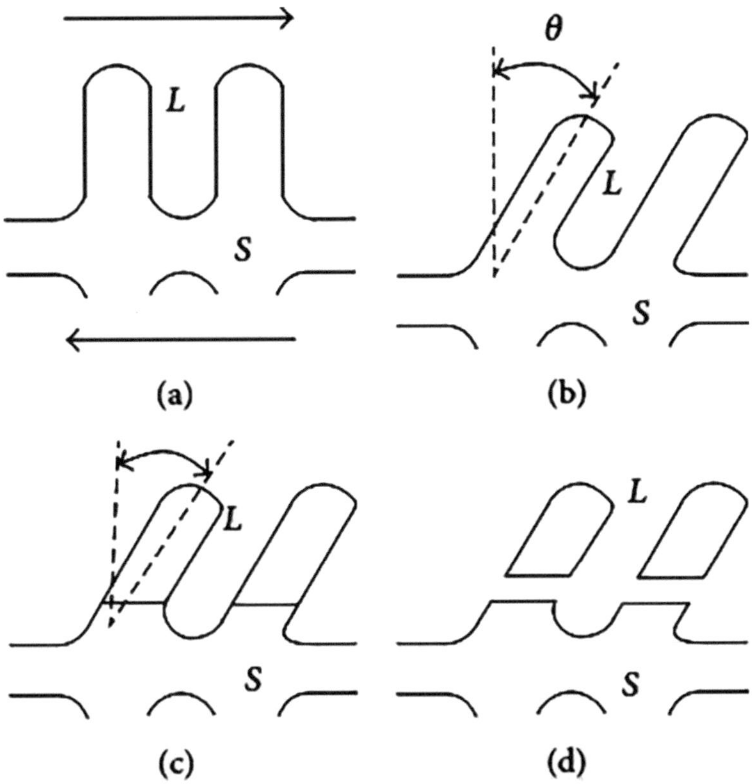

Figure 3.
Schematic illustration of the steps of the mechanism of dendrite fragmentation: (a) undeformed dendrite; (b) after bending; (c) formation of high-angle boundary; and (d) fragmentation through wetting of grain boundary by liquid metal [22].

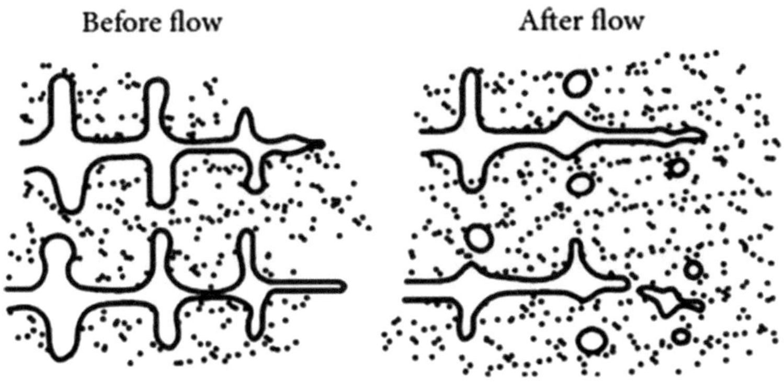

Figure 4.
Schematic diagram of dendrite multiplication theory [22].

2.2 Classification of semi solid process

The semi solid casting process mainly classified into the thixo casting and rheo casting and these processes are farther divided into many process techniques show in below (**Figure 6**).

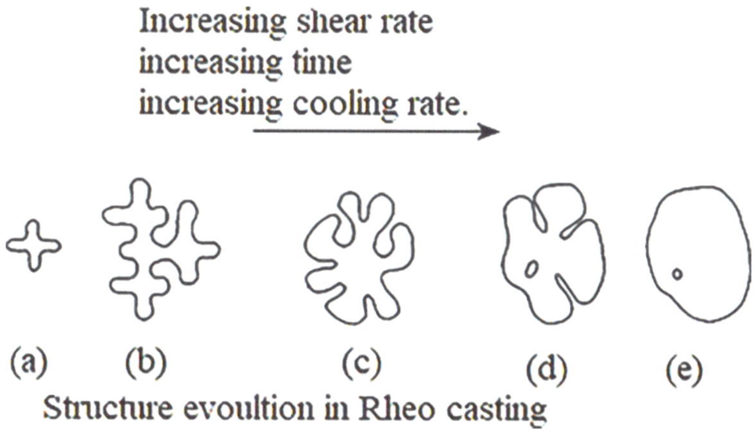

Figure 5.
Evolution of structure during solidification with shear force: (a) initial dendritic fragment; (b) dendritic growth; (c) rosette; (d) ripened rosette; and (e) spheroid [20].

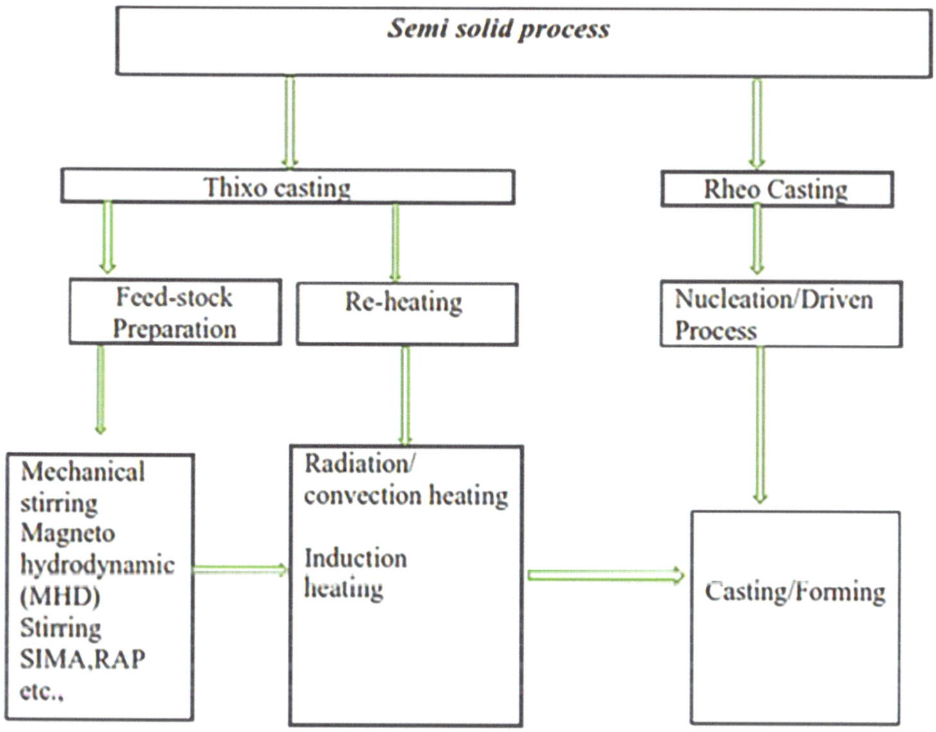

Figure 6.
Classification of semi solid processes [31].

2.2.1 Thixo casting

Thixo casting mainly consists of three separate stages the production of a pre-cast billet having the special equiaxed microstructure, the re heating of these billets to the semi-solid temperature and the casting of the components 3. Illustrated in **Figure** 7.

1. feedstock preparation;

2. Reheating of the billet; and

3. The casting process.

2.2.2 Rheo-casting

Rheo-casting is single step process to produce semi solid alloy start with liquid alloy, introduced directly into a mold without any intermediate solidification step. The semisolid slurry produced by means of different process like slope casting, new rheo casting etc. and directly introduced into a die. While thixo-forming is a route consists of reheating and forming process (**Figure 8**).

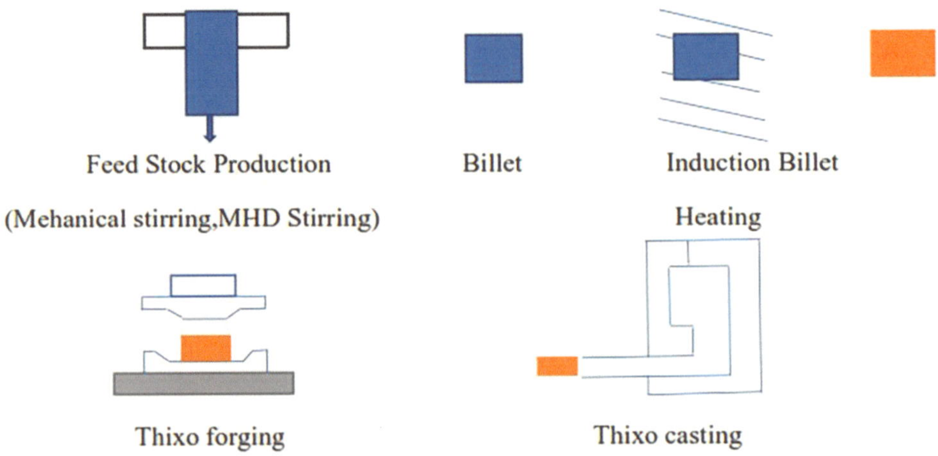

Figure 7.
Thixo casting and thixo forging [2].

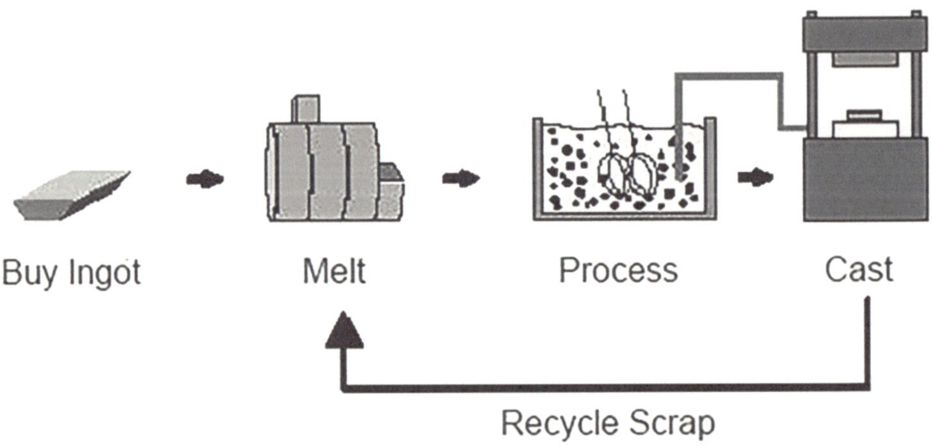

Figure 8.
Rheo casting process [2].

3. Slope casting process

Slope casting process is a rheo casting process used for the produce semi solid slurry, it consists with simple equipment and operation technique, the process carried out by pouring molten metal through channel with certain angle into a die where subsequent solidification takes place [32]. The solidification of molten alloy along a slope channel involves heat transfer, fluid flow, adhesion behavior. When the molten metal flowing through the slope channel with an angle and length [33–37], heat transfer takes between the slope channel wall and melt in contact, where generation of nuclei takes places, due to the effect of gravitation force and flow of stream the nuclei produced on slope wall are detached from the slope plate and subsequently flow through the melt stream, solid fraction of metal(semi solid slurry) observed at end of slope channel [38–42]. shear stress acting on the slurry layers and melt flow inertia restricted dendritic growth usually observed in conventional casting alloys. Illustrated in **Figure 9**. Slope casting process is a simple technique, but it can be prone to gas pick up and oxide formation which will impact negatively on mechanical properties [44, 45].

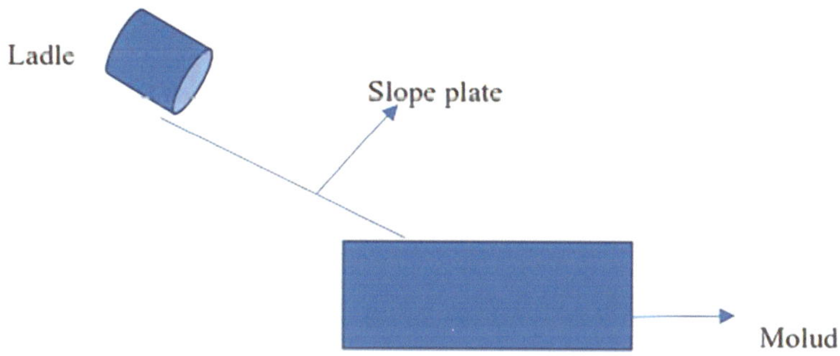

Figure 9.
Line illustration of slope casting process [43].

3.1 Mechanisms involved in slope casting process

Two mechanisms have been suggested to explain the formation of non-dendritic microstructure during flow along slope casting process. According to Haga and Kapranos et al. [46, 47], dendritic fragmentation mechanism plays an important role in slope casting process during microstructural evolution. The fragmentation of weak dendritic arms observed when the partially solidified melt collides under gravitational forces on the inclined/slope channel. Motegi et al. [48] proposed, crystal separation theory, where granular crystals nucleate and grow on the slope wall and are washed away from the wall by fluid motion illustrated in **Figure 10**.

The shear force is main factor for dendritic arm fragmentation but its effect is related to the velocity boundary layer [21] as shown in **Figure 11**.

3.2 Parameters effect the slope casting process

The process parameters in the slope casting of semisolid slurry preparation are [21, 31, 43–49]:

- Pouring temperature,

- Slope angle,
- Slope length,
- Slope plate temperature,
- Vibration of slope,
- Mold vibration etc.

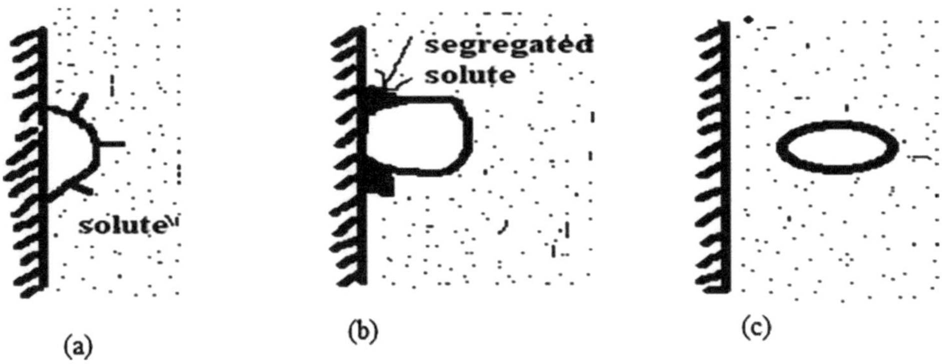

Figure 10.
Crystal separation theory (a). The generation of nuclei at slope plate wall (b). Segregation of granular crystal (c) flow through the melt [30].

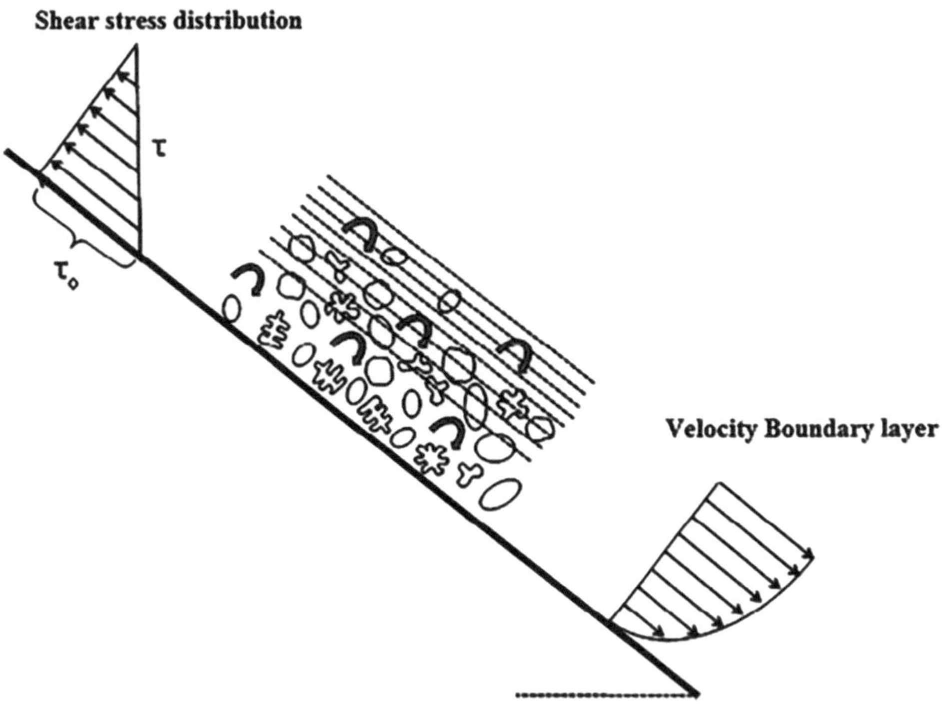

Figure 11.
Schematic diagram of the shear stress variation and velocity distribution inside the boundary layer during the flow of melt in cooling slope casting process [38].

3.2.1 Effect of pouring temperature

It is the most influencing parameter in slope casting process, T hogo et al. [36] investigated the effect of melt temperature and mold material found that pouring temperature have the great effect on the microstructure and it accounts nearly 35% of the total effect. Y Birol et al. [37] investigated the effect of pouring temperature and slope length, reported that the melt superheat required longer cooling lengths for higher pouring temperatures. Pouring with lower temperature causes formation of solid shell (formation of a thin layer of metal due to the primary nuclei that stick to the slope channel that reduces the effectiveness of the slope channel in generating nuclei) and pouring with the super-heated temperature may not get sufficient time to cool to range to produce solid nuclei on the slope plate, the main reason is that each parameter corelate each. Similar observation reported by Wen Liu et al. [39], if pouring temperature is too high a small number of primary α-aluminum phase will precipitate and some coarse primary α-aluminum phase. If the pouring temperature is too low the melt will cool rapidly and solidify. P. das et al., the temperature of the cooling plate has no prominent effect on microstructure, nevertheless a slurry with approximately 10% fraction solid can easily be obtained at the end of the plate.

3.2.2 Effect of slope length

Most of studies, slope length ranges from 200 to 800 mm, H. bidhiman et al. [41]. reported that increase in slope length that means melt flow time through channel increases it may cause the temperature drop and formation of the oxidation and solid shell as we above discussed it causes decrease in rate of heat transfer which leads to the decrease of the nucleation rate of primary solid phase, too short length does not give the proper nuclei formation and the time for the dendritic fragmentation. Slope length and slope angle are interrelated. If slope angle high need slope length should be more otherwise melt does not get sufficient time for shearing. The slope length effect on final microstructure accounts nearly 30% from studies. P. das et al. [40].

3.2.3 Effect of the slope angle

Most studies the angle ranges from 15 to 60°, the small angle is unable to give the melt to flow and shear effect on the slope plate will be less and the higher angle may cause the high velocity which does not give time to melt formation semi solid slurry and dendritic fragmentation. Farshid Taghavi and Ghassemi [42] reported the angle of slope channel had remarkable effects on the size and morphology of α-Al phase. By increasing the angle of the slope channel, the effect of shear stress and the rate of heat transfer increase. As a result, more solid particles are detached from the layer of slope channel. On the other part, duration time of shear stress and heat transfer between the melt and surface of inclined plate decrease by increase in the angle. As we above discussed in 3.2.2. the slope length and slope angle corelated to each other.

3.2.4 Effect of the slope vibration

Very few studies on effect of vibration slope on microstructural changes. Slope vibration frequency ranges from 10 to 60 Hz. Studies by Shaya Safari et al. [44],

Wen Liu et al. [39] conclude that There was no solid shell formation on the surface of slope channel by using slope vibration. The combine effect of vibration and slope channel causes increase in the amount of nucleation and nuclei due to uniform cooling rate. The mechanism in vibration slope channel is proposed that vibrating force and gravity result in Bending stress introduced in between the growing dendritic and liquid. Because of the viscous resistance of liquid, with respect to the dendritic particles and liquid phase there is a difference of the transport velocity, which causes crash among the grains and the scrub of the liquid on dendritic particles. The weak dendrite arms breakoff and form fine grains. Vibration helps the heat transfer mechanism in possible direction. The stirring caused by vibration gives rise to local temperature fluctuation of liquid phase around the primary α-al phase and Re melting of dendritic arms at the necks occurs. Which favorable to form short and homogenous small primary dendrites, equiaxed and rosette non dendritic grains.

3.2.5 Heat treatment by reheating

Researchers extended work on Slope casting process by subsequent heat treatment of casts after slope casting for better mechanical properties through spheroidization of grains and removal of defects like internal stress and porosity. Yucel Birol et al. [37] worked on the cooling slope casting and thixo forming of hypereutectic A390 alloy. Reported that The thixoformed part after slope casting process was metallurgically sound, free from porosity and revealed a uniform dispersion of fine Si particles in a homogeneous matrix. Increase mechanical properties observed. Nursen Saklakoglu et al. [33]: investigated on the microstructural evolution of ETIAL 160 aluminum feed stock produced by the cooling slope casting process experiments done with pouring temperatures of 605 and 615°C respectively subsequent isotheral heating at 565°C at 5 and 10 mins respectively, slope casting process results the primary α-aluminum dendrites has changed into α-aluminum rosette. Subsequent heat treatment helps to modify the rosette to globular structure. P das et al. [40]; too long a heating time will cause structural coarsening, while too short a heating time will lead to incomplete spheroidization of solid particles. Thus, there is a need to get optimum reheating parameters of the semi-solid alloys processed via slope casting.

3.3 Composites by slope casting

Composite materials produced using slope casting technique were reported by researchers. P. Das, [40] has studied about the semi solid microstructure of Mg_2 Si/Al composite by cooling slope casting process, reported that, the morphology of primary Mg_2Si obtained non-dendritic and size of α Al was changed to 10 from 200 μm, Toshio Haga et al. [36]. Reported that slope casting has a significant influence on the shape and grain morphology of the Metal matrix composites (MMCs). The properties of the MMCs produced by slope casting were found to be higher than those of the MMCs produced by using conventional stirring.

Distinguished the literature into table according to the optimum process parameters used in Slope Casting Process of Semi-Solid Alloys and Composites shown in **Table 1** and post parameters in **Table 2**.

Author & year	Alloy	Process parameters						
		Length of slope in (mm)	Slope angle in (degrees)	Pouring temperature in (centi grade)	Slope material, coating material and cooling medium	Slope vibration in (Hz)	Mold vibration in (Hz)	Mold material
S. R. Mukkollu et al. (2020) [43]	Al-4%cu-2%mg alloy	500	30	X	Mild steel	X	20(ultrasonic)	steel
Kerem Altug Guler et al. (2019) [35]	AA7075	650	30 & 60	660	Copper plate	X	X	Steel
N. K. Kund (2019) [21]	A356	—	—	—	—	—	—	Steel
Sahaya Safari et al. (2018) [44]	AlMg2Si	400	45	880	Copper plate, boron nitride and water	40	X	Cast-iron
Adnan Mehmood, et al., (2016) [50]	A356	800	15, 30, 45, 60 &75	800	Stainless steel and oil	×	×	Stainless steel
S. Deepak Kumar, et al. (2015) [51]	A356& A356–5TiB2	×	60	650	Mild steel and water inside	x	x	Mild steel
S. DeepakKumar et al. (2015) [52]	Al-7Si alloy	400	15, 30, 45 and 60	630, 640 and 650	water	X	X	Mild steel
Amir. A. Abdelsalam, et al. (2015) [53]	A356-Al2O3	500	60	X	low carbon steel, hard chrome and water	X	X	Steel
Saffari, et al. (2015) [54]	Al-Mg2Si	1000	45	X	Copper plate and boron nitride	40	X	X
S. Deepak Kumar, et al. (2015) [55]	A356 and A356-5TiB2	400	60	640	Mild steel & water	X	X	Mild steel

Author & year	Alloy	Process parameters						
		Length of slope in (mm)	Slope angle in (degrees)	Pouring temperature in (centi grade)	Slope material, coating material and cooling medium	Slope vibration in (Hz)	Mold vibration in (Hz)	Mold material
S. Deepakkumar, et al. (2014) [32]	A356 alloy	400	60	640	zirconia, and water	X	X	Mild steel
Amitesh Kumar, et al. (2014) [56]	High chromium cast iron	1000	15	X	Mild steel coated with graphite	X	X	Sand
Prosenjit Das, et al. (2014) [57]	A356 alloy	500	30, 60 and 45	X	Stainless steel, boron nitride and oil	X	X	Mild steel
Hamed Khosravi, et al. (2013) [58]	A356 alloy	100, 300 and 500	30, 45 and 60	660, 680 and 700	Boron nitride	X	X	Mild steel
K. S. Alhawari. (2013) [28]	A356 and Al2O3	300	60	650	Steel, boron nitride	X	X	Steel
P. Das, et al. (2013) [59]	A356 alloy	—	60 and 45	—	Boron nitride	X	X	Steel
R. Ritwik, et al. (2013) [60]	AlSi7Mg alloy	—	10	—	—	X	X	Steel
Prosenjit Das et al. (2012) [40]	A356	X	60, 45	925	Oil, boron nitride coating	X	X	Mild steel
N. Saklakoglu (2011) [34]	A380	350	60	615, 630, 650	water cooled, boron nitride	X	X	Steel
Zongning Chen (2011) [26]	Al-12Si and K2 TiF6 and KBF4	X	X	650	Water cooled	X	X	Copper
Wen Liu et al. (2011) [39]	ZAlSi9Mg	400–900	60	590–620	Copper plate and water	0–50 Hz	X	X

Author & year	Alloy	Process parameters						
		Length of slope in (mm)	Slope angle in (degrees)	Pouring temperature in (centi grade)	Slope material, coating material and cooling medium	Slope vibration in (Hz)	Mold vibration in (Hz)	Mold material
Jun X et al. (2011) [45]	A356 alloy	300, 500, 700	30, 45, 60	650, 670, 690	Mild steel and water cooled	X	X	X
H. Budiman et al. (2011) [61]	A356 alloy	250	60	610–630.	Mild steel, boron nitride	X	X	Steel
T. Haga et al. (2010) [62]	A356	30, 50, 100, 200 and 300.	15, 30, 45 and 60	620	Mild steel. Coated with BN	X	X	Steel
S. Gencalp Saklakoglu (2010) [63]	A380 alloy	500	60	630	Mild steel. Coated with BN	5.75 Hz.	X	Steel
W. Wierzchowski et al. (2010) [64]	hypoeutectic gray cast iron and high-chromium cast iron	600	0–15	TP = TL + 20 K	Copper plate, boron nitride	X	X	X
Farshid Taghavi and Ghassemi (2009) [42]	A356 alloy	20, 40, 60	20, 30, 40, 50, 60	680	Copper plate	X	X	Steel
H. Budiman et al. (2009) [41]	Al-Si alloy	250	60	620	Mild steel, boron nitride and water	X	X	Steel
Nursen Saklakoglu (2008) [33]	ETIAL 160	300	60	605, 615	Steel Plate Coated With BN And water	X	X	Steel
Yucel Birol (2008) [17]	A390 alloy	500	60	—	Steel plate and water cooled	X	X	Steel

Author & year	Alloy	Process parameters						
		Length of slope in (mm)	Slope angle in (degrees)	Pouring temperature in (centi grade)	Slope material, coating material and cooling medium	Slope vibration in (Hz)	Mold vibration in (Hz)	Mold material
Q. D. Qin and Zhao, (2007) [27]	Al/mg2Si composite	X	X	X	Alumium plate	X	X	Steel
E. Cardoso Legoretta et al. (2008) [65]	A356 alloy	150, 200	45, 60	+20 K–30 K	Mild steel coated with boron nitride and Cold water	X	X	Satinless steel
Yucel Birol (2007) [17]	A357 alloy	200, 300, 400	60	620–640	Steel plate coated with the boron nitride and water	X	X	Mild steel
Alex Muumbo et al. (2003) [66]	Cast iron	—	5–15	+20 k	Boron nitride coated	X	X	Mild, graphite, sand
Tetsuchi Motegi (2002) [48]	Al-Si-Mg alloy	80, 160, 200, 240	40, 60, 80	656, 666, 676, 686, 696	Copper plate	X	X	X
Toshio Haga (2002) [67]	A356	300	60	602, 630, 650	Mild steel coated with BN	X	X	X
Toshi Haga (2001) [36]	Al-6 Si	300	60	600	Mild steel coated with BN	X	X	Mild steel with out insulator and with insulator

Table 1.
Process parameters.

Author & year	Tensile test	Hardness	Grain size & shape factor	Findings
S. R. Mukkollu et al. (2020) [43]	x	✓	x	Refined microstructure can be obtained by if cooling slope integrated with ultrasonic acoustic cavitation.
Kerem Altug Guler et al. (2019) [35]	✓	✓	✓	Castings with the slope angle of 60 are superior to 30.
N. K. Kund (2019) [21]	✓	✓	✓	Cooling slope leads to globular and non-dendritic microstructure.
Sahaya Safari (2018) [44]	✓	✓	✓	The hardness, values of the as-cast VCS sample are higher than those of its CS
Adnan Mehmood, et al., (2016) [50]	✓	X	✓	Tensile and hardness are highest at the sloping plate at the angle of 600
S. Deepak Kumar et al. (2015) [51]	X	✓	✓	A cooling slope length of 400 mm, a low cooling slope angle of 150 was effective in dendrite fragmentation
Amir. A. Abdelsalam, et al. (2015) [53]	X	X	✓	Stir casting and cooling slope casting (SC/CSC) exhibited higher Porosity and water-cooling using SC/CSC technique effect the average size of the α-Al grains.
S. Deepak Kumar, et al. (2014) [32]	—	✓	—	Pouring temperature, which accounts for 42.08% of the total effect, followed by cooling length 40.4% and slope angle 17.44% respectively.
K. S. Alhawari. et al. (2013) [28]	—	✓	—	The hardness and wear resistance of the MMC s produced by cooling slope casting were found to be higher those of MMCs produced by using conventional stirring.
R. Ritwik, et al. (2013) [60]	—	—	—	The spheroidization effect of the alpha aluminum dendrites increases with the increase in the angle of inclination.
Prosenjit Das, et al. (2012) [40]	—	—	✓	925 K pouring temperature, 60 slope angle, 500 mm cooling length and wall temperature of 333 K has been identified as the ideal processing condition, which is in good correlation with the numerical findings
Prosenjit Das et al. (2012) [40]	X	X	✓	Spheroids and rosettes of primary Al phase has been obtained through the angle 600
N. Saklakoglu (2011) [34]	X	X	Wear test conducted	Cooling slope does not given substantial changes in friction characteristics compared to gravity casting, isothermal treatment reduced after cooling slope casting decreased the friction.
Zongning Chen (2011) [26]	✓	✓	✓	The grain size of alpha aluminum phase can be globularized using the cooling slope
Wen Liu et al. (2011) [39]	X	X	✓	Vibration effects the nucleation by increase the pressure subsequently temperature and the dendritic arms are sheared due to the vibration.
Jun X et al. (2011) [45]	X	X	X	Optimum globular microstructure with uniform distribution of A356 alloy is obtained with slope angle 45, plate length 500 mm and pouring temperature 650

Author & year	Tensile test	Hardness	Grain size & shape factor	Findings
H. Budiman et al. (2011) [41]	X	X	✓	The cooling slope casting produced smaller equiaxed α-aluminum grains with better shape factor than the conventional stirring.
T. Haga et al. (2010) [62]	X	X	X	The cooling distance affects the cooling of the melt and adhesion of solidified metal. The melt temperature becomes lower as distance becomes longer. The adhesion of the solidified metal occurs when the cooling distance becomes longer than the suitable distance.
Farshid Taghavi and Ghassemi (2009) [42]	X	X	✓	The refined and globular microstructure with a uniform reproductive distribution of A356 was obtained at an angle of 40 and a length of 40 cm
H. Budiman et al. (2009) [41]	X	X	✓	The water circulation influence on volume fraction liquid/solid grain size and shape factor.
Nursen Saklakoglu (2008) [33]	X	X	✓	Pouring at 300 mm at slope of 60 yielded more globular grains than that obtained with CS
Q. D. Qin and Zhao, (2007) [27]	X	X	✓	with increase in the isothermal holding time from 30 to 600 min the mean size of alpha-aluminum grains increases and its morphology becomes more globular
E. Cardoso Legoretta et al.(2008) [65]	X	X	X	Most of the nucleation has occurred in the upper part of the slope, the area of the impact zone plays an important role in determine the resulting microstructure and that this dominate over the cooling length
Yucel Birol (2007) [17]	X	X	X	The dissipation of the melt superheat required longer cooling lengths for higher pouring temperatures.

Table 2.
Post process parameters.

4. Conclusions

A considerable review of the literature on slope casting of semisolid Aluminum alloys suggest the following:

1. The slope casting process is a simple and cost-effective way of producing feed stock material (non-dendritic or globular) microstructure.

2. slope casting process mainly depends on the process parameters like slope length, slope angle which mainly controls the shear force on metal flow subsequently the better morphology structure obtained.

3. Reheating and isothermal holding temperature after slope casting observed better mechanical properties from different studies.

4. Using slope casting process feed stock material produced with globular microstructure is not only in cast aluminum alloys but also in aluminum metal matrix composites.

5. slope casting is best and simple process to produce the semi solid material and by using subsequent process after slope casting technique can play a prominent role in foundry industries.

6. Due to vibration on slope plate, multiple nucleations and dendritic fragmentation occur which leads to spheroidization.

Author details

Mukkollu Sambasiva Rao* and Amitesh Kumar
National Institute of Advanced Manufacturing Technology, Ranchi, India

*Address all correspondence to: samba.siva129@gmail.com

References

[1] Flemings MC. Behavior of metal alloys in the semisolid state. Metallurgical Transactions A. 1991;**22**:957-981

[2] Fan Z. Semisolid metal processing. International Materials Reviews. 2002;**47**(2):49-85

[3] Atkinson HV. Modelling the semisolid processing of metallic alloys. Progress in Materials Science. 2005;**50**:341-412

[4] Spencer DB. Ph.D. Thesis, USA: Massachusetts Institute of Technology; 1971

[5] Aguilar J, Fehlbier M, Grimmig T, Bramann H, Afrath C, Bührig-Polaczek A. Semi-Solid Processing of Metal Alloys. Steel Research International. 2004;**75**(8/9):492-505

[6] Vogel A, Doherty RD, Cantor B. Proceedings of International Conference The Solidification and Casting of Metals. Sheffield: University of Sheffield, Metals Society; 1979. pp. 518-525

[7] Kirkwood DH. Semi-solid metal processing. International Materials Reviews. 1994;**39**(5):173-189

[8] Hellawell A. Proceedings of 4th International Conference Semi-solid Processing of Alloys and Composites. Sheffield, UK: University of Sheffield; 1996. pp. 60-65

[9] Yang X, Mao W. Preparation of semisolid A356 alloy feedstock cast via a pipe consisting of partial inclined and partial vertical sections. Journal of Materials Science and Technology. 2009;**25**(2):273-276

[10] Paes M, Zoqui EJ. Semi-solid behavior of new Al-Si-Mg alloys for thixoforming. Materials Science & Engineering A. 2005;**406**:63-73

[11] Chalmers B. The structure of ingots. Journal of the Australian Institute of Metals. 1963;**8**:255-263

[12] Biloni H, Chalmers B. Origin of the equiaxed zone in small ingots. Journal of Materials Science. 1968;**3**:139-149

[13] Ohno A. Solidification: The Separation Theory and its Practical Applications. Berlin, Germany: Springer-Verlag; 1987

[14] Campbell J. Castings. Oxford: Butterworth-Heinemann; 2003

[15] Stefanescu DM. Science and Engineering of Casting Solidification. NY, USA: Kluwer Academic/Plenum Publishers; 2002

[16] Flemings MC. Solidification processing. New York: McGraw-Hill; 1974

[17] Birol Y. Cooling slope casting and thixoforming of hypereutectic A390 alloy. Journal of Materials Processing Technology. 2008;**207**:200-203

[18] Chen JY, Fan Z. Modelling of rheological behaviour of semisolid metal slurries—Part 3: Transient state behaviour. Materials Science and Technology. 2002;**18**(3):250-257

[19] Modigell M, Koke J. Rheological modelling on semi-solid metal alloys and simulation of thixocasting processes. Journal of Materials Processing Technology. 2001;**111**(1-3):53-58

[20] Atkinson HV. Rheology of Semisolid Metallic Alloys Comprehensive Materials Processing. 2014;**5**:149-161

[21] Kund NK. Cooling Slope Practice for SSF Technology. International Journal of Engineering and Advanced Technology (IJEAT). 2019;**8**(3):410-413

[22] Doherty RD, Lee H-I, Feest EA. Microstructure of stircast metals. Materials Science and Engineering. 1984;**65**(1):181-189

[23] Flemings MC, Yurko JA, Martinez RA. Solidification processes and microstructures. In: Proceedings of the TMS Annual Meeting. Charlotte, NC, USA: IIT madras; 2004. pp. 3-14

[24] Mullis AM. Growth induced dendritic bending and rosette formation during solidification in a shearing flow. Acta Materialia. 1999;**47**(6):1783-1789

[25] Zoqui JE, Robert MH. Contribution to the study of mechanisms involved in the formation of rheocast structure. Journal of Materials Processing Technology. 2001;**109**:215-219

[26] Chen Z. In situ TiB_2 Reinforced Al-Si alloy composites by semi solid processing. Material Science Forum. 2011;**675-677**:763-766

[27] Qin QD, Zhao YG. Semisolid microstructure of Mg2 Si/Al composite by cooling slope cast and its evolution during partial remelting process. Materials Science and Engineering A. 2007;**444**:99-103

[28] Alhawari KS. Wear Properties of A356/Al2O3 Metal Matrix Composites Produced by Semisolid Processing. In: The Malaysian International Tribology Conference 2013, MITC2013 Proceeding Engineering. Vol. 68. 2013. pp. 186-192

[29] Vogel A. Ph.D. Thesis. UK: University of Sussex; 1979

[30] Hellawell A. Mechanical deformation of dendrites by fluid flow. Metall. Materials Transactions. 1996;**27A**(1):229-232

[31] Mohammed MN, Omar MZ, Salleh MS, Alhawari KS, Kapranos P. Semisolid metal processing techniques for nondendritic feedstock production. The Scientific World Journal. 2013; **2013**:1-16

[32] Deepak Kumar S, Mandal A, Chakraborty M. Cooling Slope Casting Process of Semi-solid Aluminum Alloys: A Review. International Journal of Engineering Research & Technology (IJERT). 2014;**3**(7):269-282

[33] Saklakoglu N, Birol Y, Kasman S. Microstructural Evolution of ETIAL 160 Aluminium Alloy FeedstockProduced by Cooling Slope Casting. Solid State Phenomena. 2008;**141-143**:575-580

[34] Saklakoglu N, Gencalp S, Kasman S. The effects of cooling slope casting and isothermal treatment on wear behavior of A380 Alloy. Advanced Materials Research. 2011;**264-265**:42-47

[35] Keerem Altig G, Alptekin K, Ozer G, Karaaslan A. Cooling slope casting ofAA7075 alloy combined with reheating and thixo forging. Trans Non ferrous Met. soc. China. 2019;**29**:2237-2244

[36] Haga T, Suzuki S. Casting of aluminium alloy ingots for thixoforming using a cooling slope. Journal of Materials Processing Technology. 2001;**118**:169-172

[37] Birol Y. A357 thixoforming feedstock produced by cooling slope casting. Journal of Materials Processing Technology. 2007;**186**:94-101

[38] Satya SJ, Kumar V, Barekar NS, Biswas K, Dhindaw BK. Microstructural evolution under low shear rates during

Rheo processing of LM25 alloy. Journal of Materials Engineering and Performance. 2012;**21**:22

[39] Liu W, Tan J, Li J, Ding X. Influence of process papameters by vibrational cooling-shearing slope on microstructures of semi-solid ZAlSi9Mg alloy. Advanced Materials Research. 2011;**211-212**:142-146

[40] Das P, Samanta S, Venkatpathi BRK, Chattopadhyay H, Dutta P. Microstructural evolution of A356 Al alloy during flow along a cooling slope. Transactions of the Indian Institute of Metals. 2012;**65**(6):669-672

[41] Budiman H, Omar MZ, Jalar A, Syarif J, Ghazali MJ, Abdullah S. Production of feedstock material for semi-solid material processing by cooling slope casting process. International Journal of Mechanical and Materials Engineering (IJMME). 2009;**4**(2):176-180

[42] Taghavi F, Ghassemi A. Study on the effects of the length and angle of inclined plate on the thixotropic microstructure of A356 aluminum alloy. Materials and Design. 2009;**30**:1762-1767

[43] Mukkollu SR, Kumar A. Comparative study of slope casting technique in integration with ultrasonic mould vibration and conventional casting of aluminum alloy. Materials Today: Proceedings. 2020;**26**(2):1078-1081

[44] Saffari S, Akhilaghi F. Microstructure and mechanical properties of Al-Mg_2Si composite fabricated in-situ by vibrating cooling slope. Transactions of Nonferrous Metals Society of China. 2018;**28**:604-612

[45] Xu J, Wang T, Chen Z, Zhu J, Cao Z, Li T. Preparation pf semi solid A356 alloy by a cooling slope processing. Materials Science Forum. 2011;**675-677**:767-770

[46] Haga T, Kapranos P. Simple rheocasting processes. Journal of Materials Processing Technology. 2002;**130-131**:594-598

[47] Haga T, Kapranos P. Billetless simple thixoforming process. Journal of Materials Processing Technology. 2002;**130-131**:581-586

[48] Motegi T, Tanabe F, Sugiura E. Continuous casting of Semisolid Aluminum alloys. Materials Science Forum. 2002;**396-402**:203-208

[49] Motegi T, Ogawa N, Kondo K, Liu C, Aoyama S. Continuous casting of semisolid Al-Si-Mg alloy. In: Sato T, editor. Proceedings of the ICAA-6. Toyohashi; 1998. pp. 297-326

[50] Mehmood A, Shah M, Sheikh NA, Qayyum JA, Khushnood S. Grain refinement of ASTM A356 aluminum alloy using sloping plate process through gravity die casting. Alexandria Engineering Journal. 2016;**55**(3):2431-2438

[51] Deepak Kumar S, Acharya M, Mandal A, et al. Coarsening kinetics of semi-solid A356-5wt%TiB2 in situ composite. Transactions of the Indian Institute of Metals. 2015;**68**:1075-1080

[52] Acharya M, Deepak Kumar S, Mandal A. Effect of cooling slope angle on microstructure of Al-7Si alloy. Transactions of the Indian Institute of Metals. 2015;**68**:1095-1099

[53] Abdelsalam A, Mahmoud T, El-Betar A, El-Assal A. A study of microstructures characteristics of A356-Al_2O_3 composites produced by cooling slope and conventional stir cast. International Journal of Current Engineering and Technology. 2015;**15**:3560-3571

[54] Saffari S, Akhlaghi F. New semisolid casting of an Al-25Wt. % Mg_2Si composite using vibrating cooling slope. Diffusion and Defect Data Pt.B: Solid State Phenomena. 2015;**217-218**:389-396

[55] Deepak Kumar S, Mandal A, Chakraborty M. Effect of thixoforming on the microstructure and tensile properties of A356 alloy and A356-$5TiB_2$ in-situ composite. Transactions of the Indian Institute of Metals. 2015;**68**:123-130

[56] Kumar A. emi solid processing of high chromium cast iron. International Journal of Engineering Research & Technology (IJERT). 2014;**3**(5). ISSN: 2278-0181

[57] Das P, Samanta SK, Das R, Dutta P. Optimization of degree of sphericity of primary phase during cooling slope casting of A356 Al alloy: Taguchi method and regression analysis. Measurement. Volume. 2014;**55**:605-615

[58] Khosravi H, Eslami-Farsani R, Askari-Paykani M. Modeling and optimization of cooling slope process parameters for semi-solid casting of A356 Al alloy. Transactions of Nonferrous Metals Society of China. 2014;**24**(4):961-968

[59] Das P, Samanta S, Chattopadhyay H, Sharma B, Dutta P. Eulerian two-phase flow simulation and experimental validation of semisolid slurry generation process using cooling slope. Materials Science and Technology. 2013;**29**(1):83-92

[60] Ritwik R, Prasada Rao AK, Dhindaw BK. Low-convection-cooling slope cast $AlSi_7Mg$ alloy: A rheological perspective. Journal of Materials Engineering and Performance. 2013;**22**:2487-2492

[61] Budiman H, Omar MZ, Jalar A, Junaidi S. Investigation on cooling slope and conventional stir cast A356/ Al_2O_3 metal matrix composites. Advanced Materials Research. 2010;**154-155**:1284-1287

[62] Haga T, Nakamura R, Tago R, Watari H. Effects of casting factors of cooling slope on semisolid condition. Transactions of Nonferrous Metals Society of China. 2010;**20**

[63] Gencalp S, Saklakoglu N. Semisolid microstructure evolution during cooling slope casting under vibration of A380 aluminum alloy. Materials and Manufacturing Processes. 2010;**25**(9):943-947

[64] Wierzchowski W. Semi-solid processing method for cast iron. Archives of Foundry Engineering. 2010;**10**(3):149-154

[65] Legoretta EC, Atkinson HV, Jones H. Cooling slope casting to obtain thixotropic feedstock II: observations with A356 alloy. Journal of Materials Science. 2008;**43**:5456-5469

[66] Muumbo A, Takita M, Nomura H. Processing of semi solid gray cast iron using the cooling plate technique. Materials Transactions. 2003;**44**:893-900. DOI: 10.2320/matertrans.44.893

[67] Haga T. Semisolid strip casting using a twin roll caster equipped with a cooling slope. Journal of Materials Processing Technology. 2002;**130-131**:558-556

Chapter 2

Squeeze Casting Process: Trends and Opportunities

Adeolu Adesoji Adediran, A. Babafemi Ogunkola, Francis Odikpo Edoziuno, Olanrewaju Seun Adesina, M. Saravana Kumar and Osueke Christian Okechukwu

Abstract

This chapter introduces the importance of casting process, particularly in ferrous foundries. It opens with a high level functional classification of casting processes, with focus on squeeze casting, and its application in the design of metal matrix composites. To lay a suitable foundation on the subject, detailed discussions on the process parameters, process sequence, cost effectiveness, factors governing the selection of the process, associated casting defects, merits and demerits of the process are included. Special emphasis is given to discussions on the casting defects remedial measures and casting quality, types of squeeze casting processes, differences between them, area of application and components that can be manufactured using squeeze casting. The chapter closes with a brief discussion on the future trends and opportunities for improving the squeeze casting process.

Keywords: casting process, surface finish, squeeze casting, mold and foundry

1. Introduction

Metal matrix composites (MMCs) are frequently produced through squeeze casting. Because of their superior stiffness and strength than homogenous materials, MMCs are frequently used to replace engineered materials. MMCs have been used in aeronautical, automotive, and defense engineering structural applications. MMCs are typically made in industries using squeeze casting, stir casting, infiltration, and spray deposition techniques [1]. It is particularly well-known in the automotive industry for the production of diesel engine pistons. This chapter describes the process, parameters, applications and casting flaws that occurs during squeeze casting of metal matrix composites [2]. The mechanism of the defects creation is examined, as well as its implications for squeeze casting's future.

2. Casting processes

Molten materials, usually metals and their alloys, are used in casting operations. After that, the molten material is poured into a mold cavity, which takes the shape of the finished part. The molten material then cools until it solidifies into the required shape, with heat being extracted and conducted mostly through the mold. Despite the fact that the above represents a reasonably straightforward operation, casting is inherently a difficult process due to the metallurgy of working with molten metal [3].

2.1 Classification of casting processes

Casting techniques can be divided into two categories based on the type of mold.

2.2 Materials

- expendable mold processes;
- Permanent mold processes.

The molds are destroyed in expendable mold operations in order to remove the casting. Sand, plaster, and ceramics mixed with a bonding agent are common mold materials. Permanent mold procedures, on the other hand, require the mold to be created in such a way that the casting may be easily removed. Permanent molds are typically formed of metals that keep strength at high temperatures.

Various casting procedures can be used, as shown in **Figure 1**. The majority of them can handle complex geometries in a variety of weights and sizes. Overall, these casting procedures are utilized, because:

- they may make complex shapes with internal holes or hollow sections;
- several casting methods can generate extremely big components;

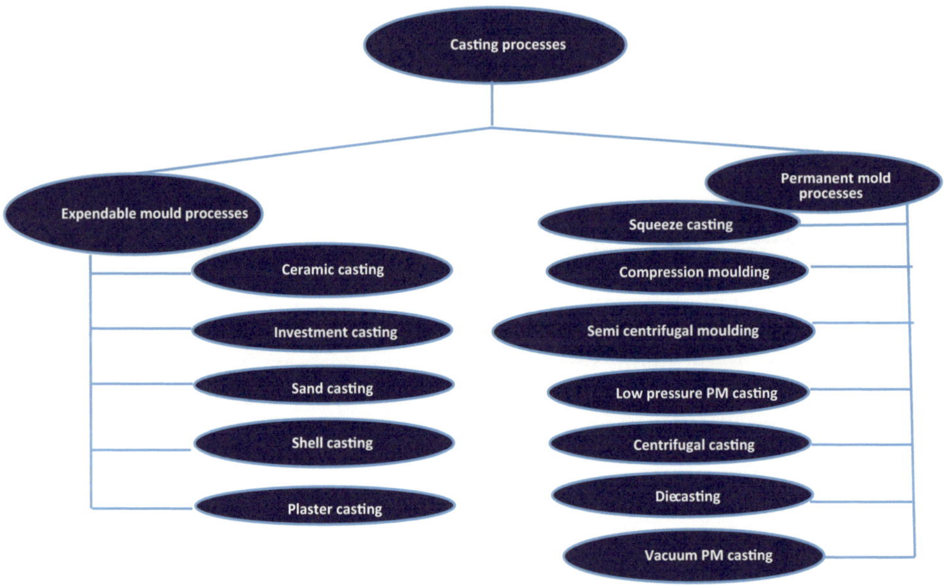

Figure 1.
General classification of casting processes [4].

- they can be used to process materials that would be difficult to treat otherwise;

- casting may be the most cost-effective way of manufacturing depending on the lot size.

Other factors are considered while determining the suitability of distinct casting techniques for a given part. Under the category of general qualities, these are discussed. Finally, casting processes are used to classify a variety of plastic processing techniques [5].

2.2.1 Importance of casting processes in ferrous foundries

1. Casting can result in products with intricate shapes and internal voids.

2. It can be utilized to make pieces that are a few hundred grams to several kilograms in weight (thousands of kilograms).

3. Any desired complicated shape can be produced.

4. Casting can be used to process both ferrous and non-ferrous materials.

5. It is cost-effective and waste-free, with excess metal from each casting being re-melted and re-used.

6. Cast metals and alloys exhibit isotropic properties. It has the same physical and mechanical qualities in both directions and along both sides.

7. Casting process is very adaptable to the needs of mass manufacturing, enabling for the rapid production of a large number of castings. The automotive industry, for example, mass-produces cast engine blocks, gearbox cases and other engine components.

8. The required casting tools are less expensive and easy to use.

9. Certain metals and alloys can only be processed through casting.

3. Squeeze casting

Because of its ability to mass produce, simpler process parameter control, improvements in wettability of the reinforcements by the liquid metal, better metallurgical quality of matrix alloys due to solidification under pressure, and the ability to reinforce only selected regions of components, squeeze casting is the preferred metal matrix composite manufacturing process for a wide range of commercial applications [6].

Squeeze casting is a hybrid of casting and forging in which molten metal is injected into a warmed die and the upper die is closed after solidification to form the mold cavity. Squeeze casting causes the metal to completely fill the cavity due to the pressure produced by the higher die, resulting in a good surface quality and low shrinkage

Both ferrous and non-ferrous alloys can be squeeze cast, however aluminum and magnesium alloys are the most frequent due to their lower melting temperatures. Parts for automobiles are a popular use [7].

The liquid metal is forced against the die walls, preventing air gaps from forming at the casting–die interface. Because pressure is applied, defects such as porosity and shrinkage are minimized, allowing for the production of finer grain castings with higher strength [8].

Preheating the die containing the preform to 300–400°C is the first step in the casting process. The punch is then driven into the die cavity at a constant ram speed of around 10 m/s after the molten metal has been injected into the die. In most circumstances, a pressure of 20–30 MPa is ideal. During solidification and a subsequent cooling time of 5–10 min, the pressure is maintained [9]. After then, the ram is removed and the composite is ejected. Squeeze casting enables for the elimination, or at least reduction, of not just gas porosity, but also flaws caused by solidification shrinkage. The origins of squeeze casting can be traced back to squeeze forming, which is a three-phased process:

1. pouring a known amount of molten metal into a pre-heated die cavity on the press's lower plate;

2. closing the die and pressurizing the liquid metal;

3. maintaining the pressure until complete solidification and extracting the casting

A schematic of a squeeze casting procedure is shown in **Figure 2**.

3.1 Types of squeeze casting process

The two basic forms of squeeze casting process may be distinguished, depending on the natural pressure applied as shown in **Figure 3**.

i. The direct squeeze casting mode.

ii. The indirect squeeze casting mode.

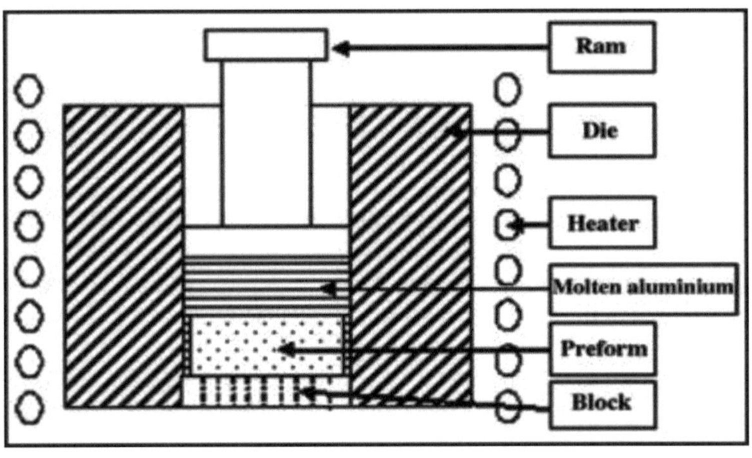

Figure 2.
Schematic representative of a typical squeeze casting machine [10].

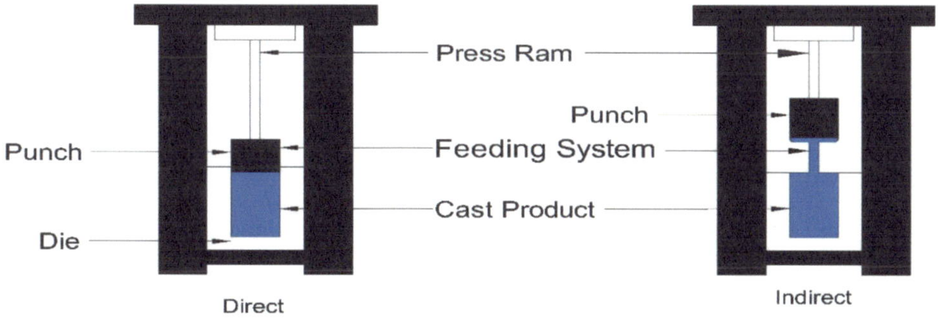

Figure 3.
Direct and indirect squeeze casting [11].

- **Direct squeeze Casting**

Direct squeeze casting (DSC) is also known as liquid metal forging. The DSC method involves Pouring liquid metal into a warmed, lubricated die and forging it while it solidifies [12]. The pressure is applied shortly after the metal begins to freeze and maintained until the entire casting has solidified. Casting ejection and handling are identical to closed die forging ejection and handling.

- **Indirect Squeeze Casting**

Direct squeeze casting (DSC) is often performed on a vertical machine (akin to a forging press), whereas indirect squeeze casting (ISC) is performed on both vertical and horizontal machines. During indirect squeeze casting, molten metal is fed to the shot sleeve and then injected into the die cavity through relatively large gates and at a low velocity (usually less than 0.5 m/s). The plunger applies high pressure "indirectly" through the huge gating system to solidify the melt in the die cavity. **Figure 4** compares the metal flow in a typical die casting method to an indirect squeeze casting method [12].

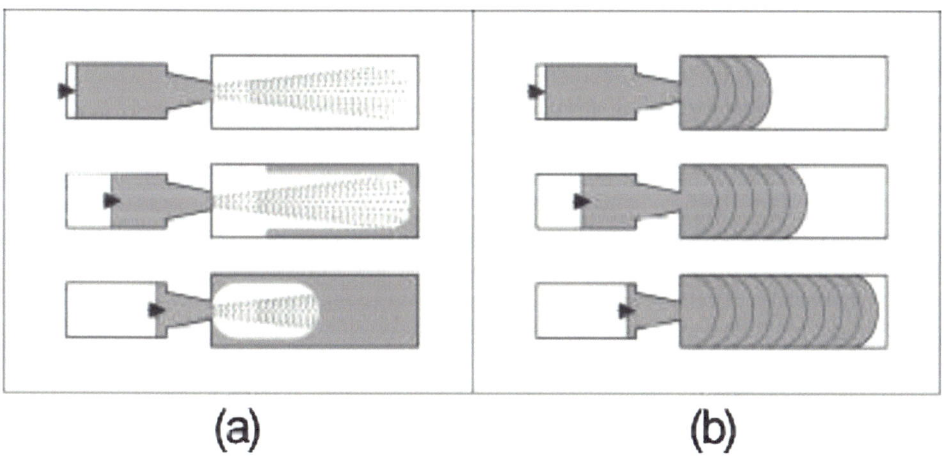

Figure 4.
Schematic illustration of metal flow in (a) conventional die casting; and (b) indirect squeeze casting process [12].

Direct squeeze casting method	Indirect squeeze casting method
The pressure for preform penetration is provided directly to the melt in the direct squeeze casting method.	The melt is forced into the preform by a gate system in indirect squeeze casting.
There is no gate mechanism, direct squeeze casting tooling is relatively easy.	The tooling is more complicated, and a gating mechanism is present.
The existence of oxide residue in the composite is another difference.	The oxide residue in the composite is stopped by the gating system.
In most cases, this is done on a vertical machine (similar to a forging press).	Indirect squeeze casting (ISC), which uses both vertical and horizontal machines, is more analogous to traditional high-pressure die casting.

Table 1.
Differences between direct squeeze casting method and indirect squeeze casting method [13].

The slower injection speed of the ISC method supports planar filling of the metal face within the die cavity, removing trapped gases from the castings (**Table 1**).

3.2 Reasons for squeeze casting

There has been a continuing need and necessity to make automobiles lighter and more fuel efficient while also improving passenger comfort. Automobile makers have been looking for solutions to keep or reduce vehicle mass. Dies have prompted die casting producers to develop new parts that were formerly iron castings or stamped steel assemblies, as well as stronger die castings that can be welded and painted. Because it gives characteristics to the metal that are difficult to accomplish with GPM casting and standard die casting, squeeze casting is typically referred to as a 'high integrity' method. Reduced porosity in the metal matrix, improved mechanical capabilities, and increased wear resistance are among the improved qualities. Squeeze castings can also be heated-treated, which is not possible with traditional die castings. In comparison to traditional die castings, thicker runner systems and, in particular, massive in-gates are utilized. The casting can be solidified under sufficient pressure to avoid practically all shrinkage by properly positioning the in-gates and maintaining high pressure on the molten alloys, as well as the use of pressure pins (if needed) during solidification. The high pressure used during solidification retains the molten metal in direct contact with the die surface, resulting in castings that are faithful to the die dimensions. Because the filling rates, also known as in-gate velocities, are low, entrapped gas in the casting is usually prevented with correct venting. As a result, the part is pore-free or substantially pore-free. Squeeze casting is a procedure that combines the benefits of both casting and forging into a single operation. The process's main selling points are the possible cost savings compared to forging and the metallurgical advantages compared to alternative manufacturing techniques. It has been demonstrated that the squeeze casting technique can generate sound and fine equiaxed grain structures in most commonly used cast alloys even some that are generally only employed in wrought form. This sound cast structure is bound to give the material isotropic properties. With excellent dimensional accuracy and repeatability, the squeeze casting technique may produce complex shapes. This allows designers to construct near-net shapes, reducing the need for further machining. Automobiles are subjected to extensive research and development in order to increase their efficiency and functionality. These enhancements frequently result in increased vehicle weight and decreased engine performance, resulting in poor fuel efficiency. Meanwhile, in order to address global environmental challenges, the desire for automobiles that are lighter and consume less gasoline is increasing. One viable option for addressing these criteria is to replace steel with aluminum. Engine blocks and gearbox cases are two examples of structural parts made from aluminum die casting. Die casting goods are currently being used in important safety elements such as suspension and space frames, which demand a high degree of strength, elongation, and yield strength.

3.3 Application of squeeze casting in the design of metal matrix composites (MMC)

Depending on the type of hardening particles used and the intended application condition, metal matrix composite materials are made via casting or powder metallurgy. Aluminum, magnesium, copper, titanium alloys, and super alloys are the most frequent metal matrixes [14]. Graphite, carbon, oxides, carbides, boron, molybdenum, and tungsten are common hardening particle or fiber materials. Casting route is used to make the majority of reinforced aluminum. Depending on the composition of the aluminum alloys, several casting procedures are used. The most common casting methods are gravity, vacuum, rotary centrifugal, squeeze, and extrusion casting. In some cases, an automatic bottom pouring stir casting furnace (**Figure 5**) for melting aluminum and/or magnesium metals and alloys is provided, along with the various casting methods. The bottom pouring stir casting furnace has a bottom furnace that pours molten metal composites directly into the casting equipment by activating a single mechanism while the stirring process continues [16].

As a result, new casting procedures have been developed to minimize these flaws. Squeeze casting, out of all the current casting processes, offers the greatest potential

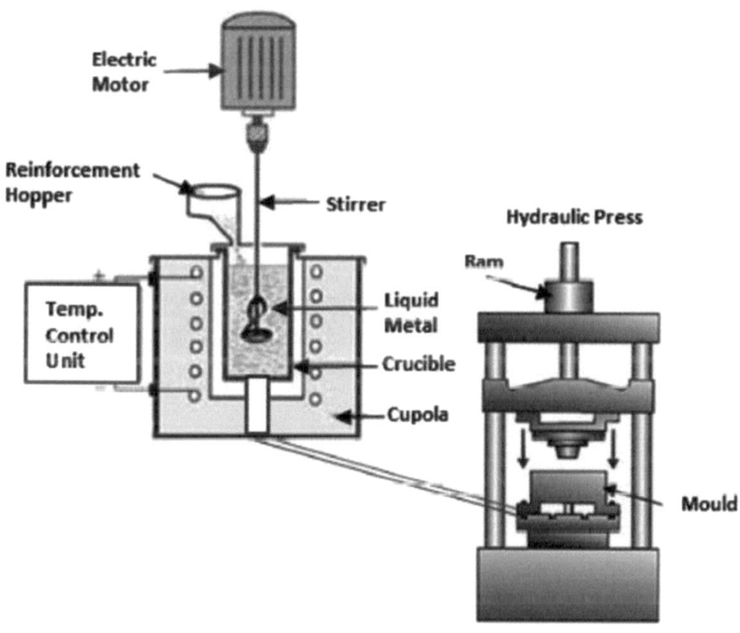

Figure 5.
Bottom type stir casting set up with squeeze casting attachment [15].

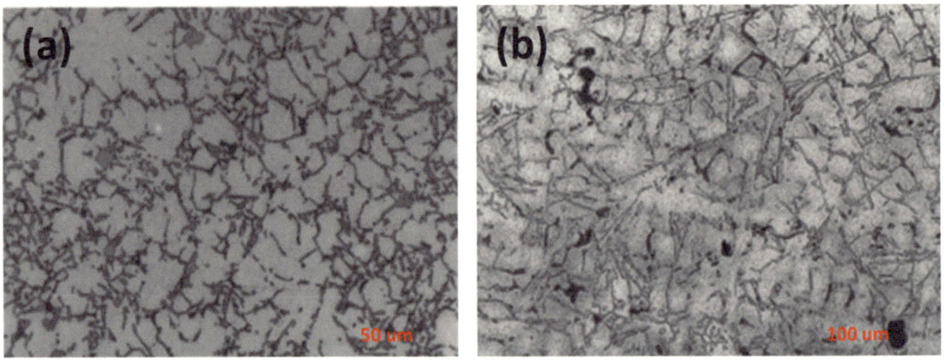

Figure 6.
A micrograph of (a) squeeze casting and (b) conventional casting [17].

Process parameter	Influence if increased
Melt temperature	Improved fluidity, dissolved gas content, and vulnerability to hot tears
Die temperature	Solidification is slowed, and tolerances are altered
Punch speed	Faster die filling means a greater possibility of jetting
Pressure	Better consolidation, fewer flaws, and a higher risk of seizing and galling
Pressure hold time	Interfacial bonding between different materials is improved
Lubrication thickness	Better fluidity, lubricant may be embedded into the part surface (poor surface quality), and lubricant buildup can jam moving die components, particularly ejectors

Table 2.
Squeeze casting factors that have an impact on part quality [9].

for producing fewer defective cast components. **Figure 6** shows the micrograph of a) squeeze casting and b) conventional casting (**Table 2**).

3.4 Squeeze casting process sequence

1. A quantity of molten metal is poured into a warmed die cavity on a hydraulic press's bed.

2. The press is turned on to pressurize the liquid metal and shut the die cavity. This is the process of molten metal solidifying under pressure.

3. The metal is held under pressure until it has solidified completely. Not only does this boost the rate of heat flow, but it also has the potential to remove macro/micro shrinkage porosity. Furthermore, because gas porosity nucleation is pressure-dependent, porosity formation due to dissolving gases in molten metal is limited.

4. The punch is finally removed, and the component is ejected.

The squeeze casting sequence of operation is schematically illustrated in **Figure 7.**

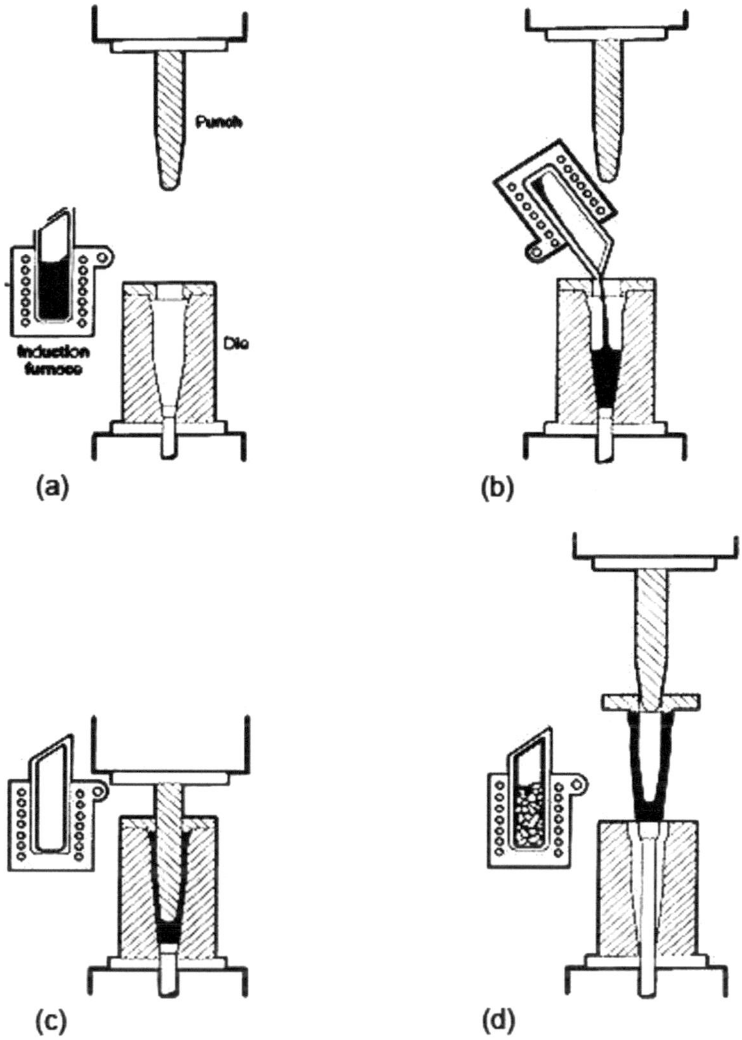

Figure 7.
Schematic illustrating squeeze casting process sequence of operations. (a) Melt charge, preheat, and lubricate tooling. (b) Transfer of melts into die cavity. (c) Close tooling, solidify melt under pressure. (d) Eject casting, clean dies, charge melt stock [18].

3.5 Process parameters

The primary process parameters during squeeze casting are as follows:

1. The temperature of molten metal varies depending on the alloy and part geometry. The starting point is usually 35–55°C above the melting point.

2. Die temperatures typically range from 300 to 400 degrees Celsius.

3. Squeeze pressures of 20–30 MPa are commonly utilized, with 25 MPa being the most common, depending on the part geometry and mechanical qualities required.

4. Pressure duration: can range from 30 to 120 seconds, depending on part geometry.

5. Lubrication level: When sprayed on the warm dies before to casting, a suitable grade of graphite spray lubricant has proven sufficient for aluminum, magnesium, and copper alloys. To prevent welding between the casting and the metal die surfaces, ceramic-type coatings are necessary.

6. Punch to-die-clearance button metal volume

7. Quality and quantity of the melted product

Process parameters: a number of parameters that can affect the casting quality for both direct and indirect Squeeze casting are explained below;

- The first is the alloy and its quality; in reality, the alloy's melting temperature and thermal conductivity influence die life and dictate casting parameters like die temperature. As a result, squeeze casting is preferred for low melting temperature alloys like Al and Mg. Metal cleaning is also crucial to avoid dross and oxide impurities in the casting.

- As previously stated, melt quantity is critical in direct SC, and precision control systems are required to ensure casting dimensional control. The die cavity can be built to accommodate the presence of an enlarged appendix in a noncritical area for the distribution of any extra metal. Lynch offered a compensating hydraulic piston and cylinder to control the exact amount of metal in the die as an alternative. Overflows are also a viable option.

- The heat transfer rate and alloy cooling are affected by the operating temperature of the die cavity and punch. In reality, too low a temperature in the die can produce premature solidification and cold laps in the casting, while too high a temperature in the die can create surface flaws and metallization (casting and die welding). For Al and Mg alloys, the die temperature is normally between 200 and 300°C, with the lower temperature being ideal for thicker section parts. A lubricating agent, usually made of graphite, should be employed.

- The time between the actual pouring of liquid metal and the instant the punch begins forcing the alloy into the die cavity is known as the time delay. A time delay is proposed to allow cooling of the metal pool before squeezing in order to prevent shrinkage porosity. This time varies based on the melt/pouring temperature and the casting's complexity. Two additional important parameters are the magnitude and duration of the applied pressure. Pressure has a direct impact on the microstructure and mechanical qualities of squeeze cast components because it determines the solidification temperature.

Squeeze casting is simple and low-cost, makes good use of raw materials, and has a lot of potential for automation at high speeds. The technique produces cast products with the best mechanical qualities possible. A fluid metal is solidified under pressure during solidification in the squeeze casting process, resulting in a high cooling rate and temperature gradient. Squeeze casting has a lot of advantages, including low porosity density, heat treatability, consistency, and good mechanical qualities [11].

3.6 Merits and demerits of squeeze casting

Merits

a. Its mechanical qualities are superior.

b. It has a finer structure and less porosity.

c. It has a smoother, better-finishing surface.

d. It is extremely profitable.

e. Possibility of use in composite goods.

f. Post-casting machining is minimal or non-existent.

Demerits

a. The metallic mold's life expectancy is reduced.

b. High precision control is required.

c. Because of the intricate tooling, costs are extremely expensive.

d. Tooling has no versatility.

3.7 Associated casting defects

I. **Oxide inclusions:** Failure to maintain clean melt-handling and melt-transfer systems is the cause of oxide inclusions. Filters should be included in the melt-transfer system, or molten metal turbulence should be controlled when filling the die cavity, to reduce the chances of introducing metallic inclusions. It's also a good idea to keep foreign materials out of open dies.

The image below represents oxide inclusion (**Figure 8**).

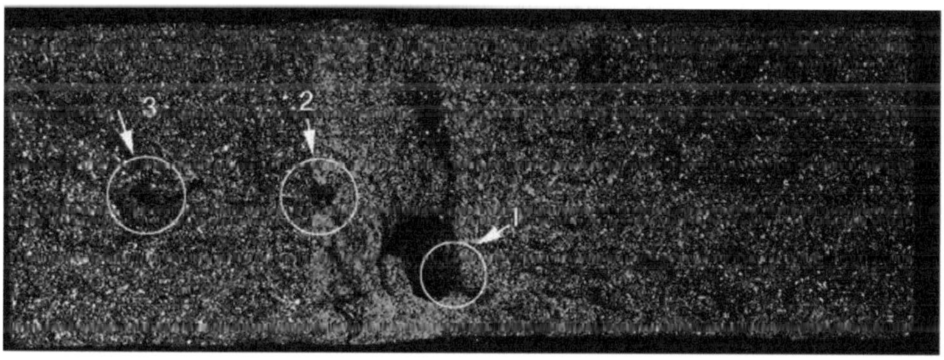

Figure 8.
Oxide inclusion on materials [19].

II. **Porosity and voids:** When not enough pressure is applied during squeeze casting procedures, this can happen. When the other factors are tuned, porosity and/or voids are usually reduced by increasing the casting pressure. The image below represents porosity on cast materials (**Figure 9**).

III. **Extrusion Segregation:** Squeeze cast components have significantly less relative micro segregation than other cast components. Defects like these can be avoided by properly designing dies, using a multiple gate system, increasing die temperature, or reducing the delay time before die closure (**Figure 10**).

IV. **Centerline segregation:** At lower solute temperatures, it's a fault that is commonly found in high-alloy wrought aluminum alloys. As the lower-melting solute is contained within the center sections of the extruded projections or more massive areas of the casting as solidification occurs on the die walls, the liquid phase becomes

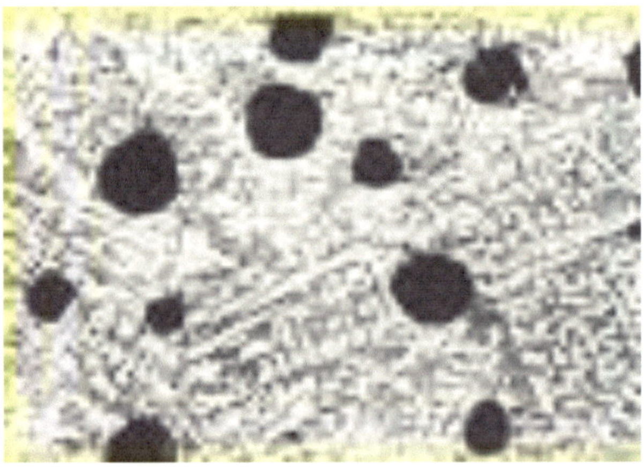

Figure 9.
Porosity on cast materials [20].

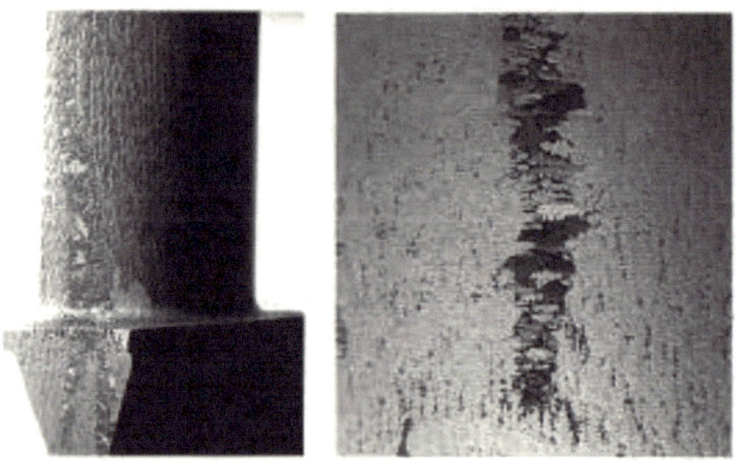

Figure 10.
Image of segregation on cast materials [21].

more concentrated. Such flaws can be eliminated by increasing die temperature, reducing die closure time, or choosing a different alloy (**Figure 11**) [18].

V. **Blistering:** During turbulent die filling, trapped air or gas from the melt generates blisters on the cast surface when the pressure is released or during subsequent solution heat treatments. Degassing the melt and preheating the handling transfer equipment, utilizing a slower die closing speed, increasing the die and punch venting, and lowering the pouring temperature are all ways to avoid such problems (**Figure 12**).

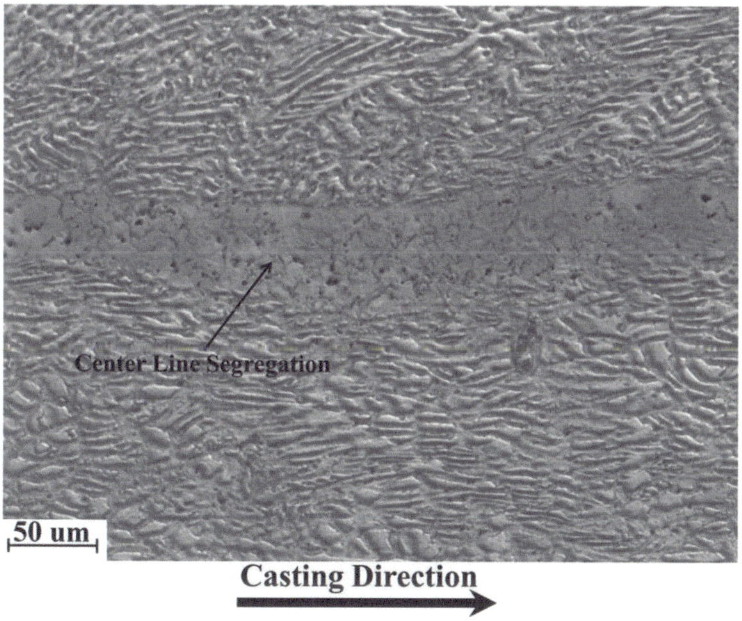

Figure 11.
Optical image of centerline segregation [22].

Figure 12.
Image of blistering on metal [23].

VI. **Cold laps:** Molten metal covering previously solidified layers causes inadequate bonding between the two. It's vital to raise the pouring temperature or the die temperature to avoid chilly laps. It has also been discovered that reducing the die closure time is useful (**Figure 13**).

VII. **Hot tearing:** occurs in alloys with a wide freezing temperature range. Contraction of the solid around the stiff mold surface can cause rupture in partially solidified portions when solid and liquid coexist over a wide range of temperatures. Reduced pouring temperature, reduced die temperature, increased pressurization duration, and increased draft angles on the casting are some of the strategies utilized to avoid hot ripping in squeeze cast goods (**Figure 14**).

VIII. **Sticking.** Rapid cycling of the process without proper die/punch cooling and lubrication causes a thin layer of casting skin to adhere to the die surface. Reduce the temperature of the die or the temperature of the pouring liquid to avoid sticking (**Figure 15**).

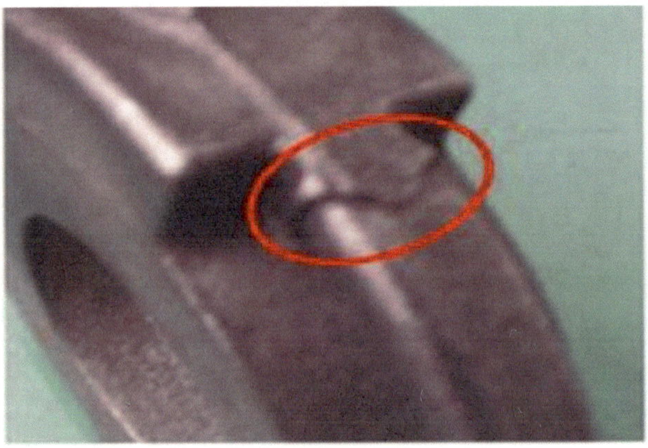

Figure 13.
Image of cold lapping of a metal represented below [24].

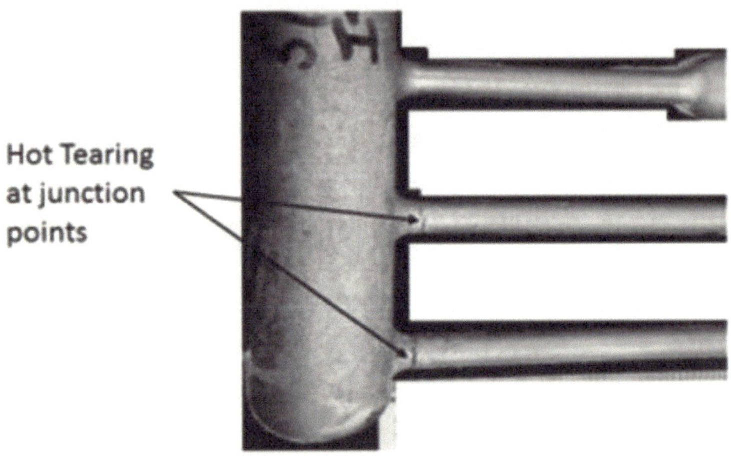

Figure 14.
Hot tearing shown on a cast metal [25].

Figure 15.
Image of stcking [26].

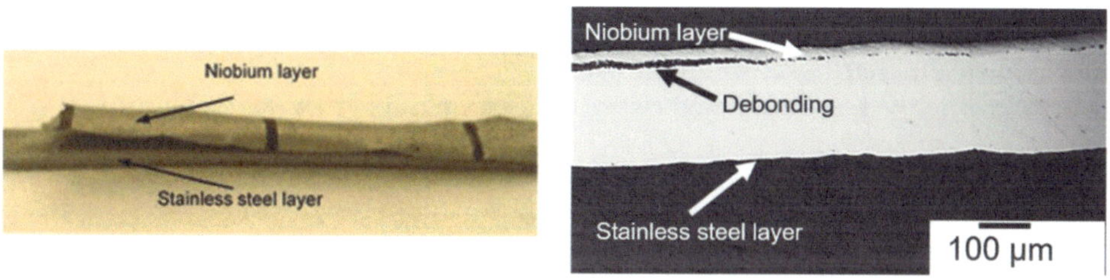

(a) (b)

Figure 16.
Image of extrusion on a metal [27].

IX. **Extrusion debonding** When the metal sits in the open die for a long time before being extruded to fill the die cavity, this happens. After the melt has been extruded around the partially hardened crust in the die, the oxide remains there, resulting in the absence of a metal-to-metal link at oxide stringer positions. Increase the tooling or pouring temperatures to prevent extrusion debonding. The production of oxide on the semi-liquid metal in the die can be reduced by reducing the die closure time (**Figure 16**) [28].

4. Future trends and oppportunities for improving squeeze casting process

The introduction of squeeze casting as a production technique has provided a solution to these criteria, with the present emphasis on lowering materials use through near-net shape processing and the necessity for both higher strength and high ductility parts. Squeeze casting has been used in manufacturing in the United States, the United Kingdom, and Japan in recent years. Squeeze casting is gaining popularity in the industry. Squeeze casting was first advertised as a solution for components that had difficulty with traditional castings. This could have been due to design constraints, new applications resulting in higher loads, increasing pressure tightness requirements, or a desire to improve customer reliability. Following this initial surge in the industry, replacements for aluminum forgings and conversions from ferrous castings, such as ductile iron, were the following growth areas. While smaller tonnage machines are frequently utilized in Japan, where customers are more diverse, squeeze castings are mostly used in the automotive industry in the United States. These uses are driven by the need to reduce vehicle weight. This necessitates the conversion of iron to aluminum, as well as the requirement for cost savings, which necessitates the conversion of existing processes such as forging and permanent mold to new ones. The majority of the automotive items are steering and suspension components, such as steering knuckles and control arms, as well as air conditioning components. Squeeze casting is the recommended method when fatigue and ductility are critical component qualities. There is a significant chance for squeeze castings to flourish as industry expertise with squeeze casting is combined with technical understanding. In the past, the method has been used on defective parts or poorly designed designs, with disastrous outcomes. However, when a deeper understanding of the process' capabilities is gained, and technically competent models and algorithms for squeeze casting optimization are established, new applications will consistently meet or surpass expectations. Squeeze casting parts and components can be used to replace cast iron and pressed steel parts in automobiles. Furthermore, when the pressure increases, structural improvements occur, allowing the tensile characteristics to be comparable to wrought aluminum [29].

4.1 Some components produced by squeeze casting

The squeeze casting process has a number of applications which include;

Dome, blades, disks, automotive wheels, pistons, gears, hydraulic brake valve, Brake master cylinder, steering knuckles, control arms.

5. Conclusion

The process sequence, parameters and properties of squeeze casting operation with the attendant technical advantages over other conventional casting production techniques have been discussed in the foregoing sections. Squeeze casting of metal matrix composites is amenable to mass production, especially in the automotive industry. The process and production sequences can be subjected to automatic process control with the gains of consistent high casting quality and production rates. There exist a plethora of opportunities for future improvement and optimization of squeeze casting process parameters as researches and development efforts in the automobile industry and metal matrix composites is intensified.

Author details

Adeolu Adesoji Adediran[1], A. Babafemi Ogunkola[1*], Francis Odikpo Edoziuno[2], Olanrewaju Seun Adesina[3], M. Saravana Kumar[4] and Osueke Christian Okechukwu[3]

1 Department of Mechanical Engineering, Landmark University, Omu-Aran, Kwara State, Nigeria

2 Department of Metallurgical Engineering, Delta State Polytechnic, Ogwashi-Uku, Nigeria

3 Department of Mechatronics, Bowen University, Osun State, Nigeria

4 Department of Production Engineering, National Institute of Technology, Tiruchirappalli, Tamil Nadu, India

*Address all correspondence to: ogunkolababafemi111@gmail.com

References

[1] Paramasivam K, Vijay Anand M, Sambathkumar M. Investigation of optimum process parameter of lost foam casting of A356/SiC metal matrix composite. Materials Today: Proceedings. 2021;**47**. DOI: 10.1016/j.matpr.2021.06.035

[2] Papworth A, Fox P. Oxide film casting defects in squeeze cast metal matrix composites. Materials Letters. 1996;**29**(4-6):209-213. DOI: 10.1016/S0167-577X(96)00148-6

[3] Kapranos P, Carney C, Pola A, Jolly M. In: Hashmi S, Batalha GF, Van Tyne CJ, Yilbas BBT-CMP, editors. Advanced Casting Methodologies: Investment Casting, Centrifugal Casting, Squeeze Casting, Metal Spinning, and Batch Casting. Oxford: Elsevier; 2014. pp. 39-67.

[4] Mital A, Desai A, Subramanian A, Mital A. Selection of manufacturing processes and design considerations. Product Development. 2014:133-158. DOI: 10.1016/B978-0-12-799945-6.00006-5

[5] Scallan P. In: Scallan PBT-PP, editor. Material Evaluation and Process Selection. Oxford: Butterworth-Heinemann; 2003. pp. 109-170

[6] Senthil P, Amirthagadeswaran KS. Optimization of squeeze casting parameters for non symmetrical AC2A aluminium alloy castings through Taguchi method. Journal of Mechanical Science and Technology. 2012;**26**(4):1141-1147. DOI: 10.1007/s12206-012-0215-z

[7] Vijian P, Arunachalam VP. Optimization of squeeze casting process parameters using Taguchi analysis. International Journal of Advanced Manufacturing Technology. 2007;**33**(11):1122-1127. DOI: 10.1007/s00170-006-0550-2

[8] Chelladurai SJS, Arthanari R, Nithyanandam N, Rajendran K, Radhakrishnan KK. Investigation of mechanical properties and dry sliding wear behaviour of squeeze cast LM6 aluminium alloy reinforced with copper coated short steel fibers. Transactions of the Indian Institute of Metals. 2018;**71**(4):813-822. DOI: 10.1007/s12666-017-1258-8

[9] "lectureofsqueezecastingosama2 (1)"

[10] Casting D, Composite M, Composite MM, Oxide A, Material M. Fundamentals of Metal Matrix Composites Casting Routes for Production of Metallic Based Composite Parts Corrosion of Metal Matrix Composites Applications: Magnesium-based metal matrix composites (MMCs) Squeeze Casting for the Production of Metallic. 2021. pp. 2016-2018

[11] Ghomashchi MR, Vikhrov A. Squeeze casting: An overview. Journal of Materials Processing Technology. 2000;**101**(1):1-9. DOI: 10.1016/S0924-0136(99)00291-5

[12] Vinarcik EJ. Understanding defects in high integrity die castings. SAE Transactions. 2003;**112**:405-413

[13] Kwok TWJ, Zhai W, Peh WY, Gupta M, Fu MW, Chua BW. Squeeze casting for the production of metallic parts and structures. Encyclopedia of Materials: Metals and Alloy. 2021:87-99. DOI: 10.1016/B978-0-12-819726-4.00038-7

[14] Kapranos P, Carney C, Pola A, Jolly M. Advanced casting methodologies: Investment casting, centrifugal casting,

squeeze casting, metal spinning, and batch casting. Comprehensive Materials Processing. 2014;**5**:39-67. DOI: 10.1016/B978-0-08-096532-1.00539-2

[15] Kannan C, Ramanujam R. Comparative study on the mechanical and microstructural characterisation of AA 7075 nano and hybrid nanocomposites produced by stir and squeeze casting. Journal of Advanced Research. 2017;**8**(4):309-319. DOI: 10.1016/j.jare.2017.02.005

[16] Bottom Pouring Type Stir Casting Furnace to form Metal Matrix. Available from: https://www.quickcompany.in/patents/bottom-pouring-type-stir-casting-furnace-to-form-metal-matrix-composite-using-aluminium-magnesium [Accessed: January 25, 2022]

[17] Verma SK, Dorcic JL. Squeeze casting process for metal-ceramic composites. SAE Transactions. 1987;**96**:143-154

[18] Squeeze casting process: Part One: Total Materia Article. Avialble from: https://www.totalmateria.com/page.aspx?ID=CheckArticle&site=ktn&LN=TH&NM=172 [Accessed: January 25, 2022]

[19] Oxide Inclusion. Available from: https://www.giessereilexikon.com/en/foundry-lexicon/Encyclopedia/show/oxide-inclusion-3936/?cHash=ade0f10720424c9d84b2636a8006a316 [Accessed: February 05, 2022]

[20] Casting Defects Intoduction Casting is Defined as Something. Available from: https://slidetodoc.com/casting-defects-intoduction-casting-is-defined-as-something/ [Accessed: February 06, 2022]

[21] Beckermann C. Examples of macrosegregation in alloy casting 1.1. Materials Science and Technology. 1974:4733-4739

[22] Optical Image of the Center-line Segregation, the Microstructure of the... | Download Scientific Diagram. Available from: https://www.researchgate.net/figure/Optical-image-of-the-center-line-segregation-the-microstructure-of-the-segregate-is_fig4_272640074 [Accessed: February 06, 2022]

[23] Plated Surfaces should be Blister Tested if Brazing is Involved (Part 1) | 2019-02-21 | Industrial Heating. Available from: https://www.industrialheating.com/blogs/14-industrial-heating-experts-speak-blog/post/94782-plated-surfaces-should-be-blister-tested-if-brazing-is-involved-part-1 [Accessed: February 06, 2022]

[24] & 5: Examples of Cold Shut [12] | Download Scientific Diagram. Available from: https://www.researchgate.net/figure/5-Examples-of-Cold-Shut-12_fig3_300258417 [Accessed: February 06, 2022]

[25] Hot Tearing of Aluminum Alloys: Total Materia Article. Available from: https://www.totalmateria.com/page.aspx?ID=CheckArticle&site=ktn&NM=406 [Accessed: February 06, 2022]

[26] Defect that Casting Stick to Tooling | Reason and solution. Available from: https://www.prodiecasting.com/reason-and-solution-die-casting-parts-stick-to-cavity-of-tooling/ [Accessed: February 06, 2022]

[27] Khan HA, Asim K, Akram F, Hameed A, Khan A, Mansoor B. Roll bonding processes: State-of-the-art and future perspectives. Metals. 2021;**11**(9):1344

[28] Squeeze Casting Process: Part two: Total Materia Article. Available from: https://www.totalmateria.com/page.aspx?ID=CheckArticle&site=ktn&LN=EN&NM=174 [Accessed: January 27, 2022]

[29] Iyer A. Squeeze casting: The future. International Specifications of Ski Instructions. 2011:1-48

Section 2

Green Sand Molding and Molten Metal Flow in Molds

Chapter 3

Methods to Determine, Influence and Improve the Flowability of Sand Mixtures

Vincent Haanappel

Abstract

Due to increasing environmental awareness, more and more foundries started to replace the organic binders with an environment-friendly binder system based on inorganics. However, inorganic binder systems can only replace the conventional organic binders if all requirements, set by the foundry industry are being fulfilled. One of these is the production of high-quality cores, indicative of good compaction, no surface defects, and high surface smoothness. Such types of cores can only be manufactured when the sand mixture shows sufficient flowability. This chapter presents a study on the flowability of various types of sand mixtures all including inorganic binder systems finally to be used for the production of sand cores for the foundry industry. Results have shown that the flowability of sand mixtures can be modified and improved by (1) the addition of small amounts of surface-active agents and/or (2) well-chosen additives characterized by micro-sized spherical particles. The addition of only a few amounts of surface-active agents resulted in a significant improvement of flowability and thus of core quality. Similar results were achieved with the use of small concentrations of spherical micro-sized particles.

Keywords: flowability, sand mixture, inorganic binder, alkali silicate, surfactant, surface tension, additive, spherical particles, powder flow tester, cores

1. Introduction

Environment-friendly binder systems for molds and cores are generally based on the use of a 2-component inorganic binder system consisting of a liquid alkali silicate and a powder-like solid consisting of various minerals. Since environmental awareness receives today more and more attention, the replacement of organic by inorganic binders result can be considered in this, resulting in a significant reduction of dirty smell and toxic emissions. This means that the introduction of the so-called inorganic binder system needs many research and development activities before it can replace the old-fashion organic binder systems. In this context, with the existing technology to heat cure the cores, high strength can be achieved, comparable with that obtained with the existing organic binders.

The new environment-friendly binder system SOLOSIL TX developed by Foseco, Vesuvius, is an inorganic thermosetting core binder, which combines the advantages of the organic binder regarding strength and productivity with environmental friendliness [1].

Achieving high-quality castings with the use of cores including the above-mentioned inorganic binder systems depend not only on the casting process itself and their processing parameters, but also on the quality of these cores. Use of cores with insufficient strength or with locally low compaction results in lower surface smoothness and often in defects of the casting surface. The main requirements to achieve high-quality cores and received from the foundry industry are:

- Use of relatively cost-effective and environment-friendly materials
- Easy to handle and to mix with various types of sand
- Sufficient bench life of the sand mixture
- Good flowability of the sand mixture
- No blocking of the inlet shooting nozzles
- High-quality cores with sufficient compaction and surface smoothness
- Short cycle times = short core production time
- Sufficient handling and casting strength of cores
- Sufficient breakdown/de-coring properties
- No gas formation during the casting process resulting in bad castings
- To achieve sufficient storage stability of the produced cores.

It is not the aim of this chapter to highlight and/or to discuss all parameters affecting the manufactured castings; this contribution is dealing with only one aspect: how to produce good quality cores (read: high compaction, no surface defects and high surface smoothness) with quartz sand as the filling material and well-chosen inorganic binder systems, the latter to achieve an environment-friendly working area.

High-quality cores can only be produced when the flowability of the sand mixture is such that the core box is filled completely. This means that the surface friction between individual sand grains and the wall friction, meaning the friction between the sand mixture and the surface of the core box, should be sufficiently low. **Figure 1** shows a small part of two types of cores, one (left picture) with insufficient compaction due to low flowability, and the other (right picture) with high compaction—high flowability of the sand mixture.

Flowability of sand mixtures depends on various parameters: the nature and amount of the liquid binder, the nature and amount of the optional additive(s), and the type of the filling material. Regarding the additive and filling material, the flow properties of these components depend on several parameters, such as particle size distribution, particle shape, chemical composition, moisture and temperature [2].

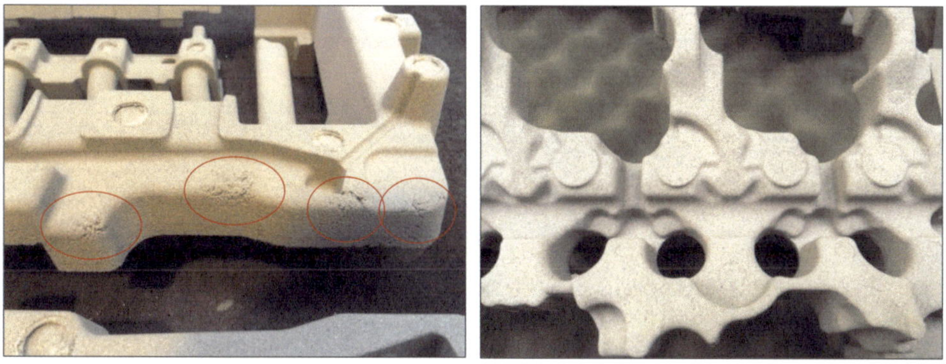

Figure 1.
Left: core with locally insufficient compaction (low flowability of the sand mixture); right: core with high compaction (high flowability of the sand mixture). Both cores were manufactured with inorganic binder systems.

Generally, the filling material is a given since foundries obtained their sand from local sand quarries. This means that only the nature and properties of the liquid binder and/or additive are (is) accessible for modification.

The objective of this chapter is to highlight some methods to determine, to influence and to optimize the physical and chemical properties of inorganic binder systems aiming at high flowability of the sand mixture.

2. Experimental

2.1 Equipment

2.1.1 Attension optical tensiometer

With the Attension optical tensiometer surface tension of liquids can be analyzed following the pendant drop method. Since the drop shape is partly related to the surface tension of the liquid, the captured image will be analyzed with a drop profile fitting method calculating the surface tension. This instrument has already shown its high accuracy and fast sampling rate, so results can be obtained within a few seconds. **Figure 2** shows some pictures of this tensiometer.

2.1.2 PFT powder flow tester

The powder flow tester was initially developed to characterize the flow behavior of solid powder material with particle sizes up to a maximum of about 1 mm. Since there was also a need to determine and to define the flowability of sand mixtures with a relatively small amount of a liquid, it was considered to use the PFT also for these applications. To compare different types of sand mixtures, results will be published by a so-called "flow function" plot. This flow function plot shows the flowability of various types of samples over so-called different "consolidation stresses", the latter being considered as compressive stress. This plot shows various regions starting from free-flowing, to easy flowing, cohesive, very cohesive to non-flowing. The lower the curve the higher the measured flowability. **Figure 3** shows the PFT Brookfield powder

Figure 2.
Attension optical tensiometer.

Figure 3.
PFT powder flow tester to measure flowability of powders and sand mixtures.

flow tester. About this equipment, the weight of the sand mixture placed in the sample holder, can also be an indirect indication of the flowability. In this case, the Hausner ratio (Eq. (1)) [3] can be used to obtain a more quantitative value of the flowability:

$$H = \rho_T/\rho_B \tag{1}$$

where ρ_T can be considered as the bulk density under pressure and ρ_B the freely settled bulk density of the mix. The Hausner ratio is not an absolute value for flowability but gives only a quantitative approach. The higher the Hausner ratio, the lower the flowability. In the industry, also the Carr index C (Eq. (2)) [4] is sometimes used, which is another value of flowability but related to the Hausner ratio:

$$C = 100 - 100/H \tag{2}$$

where H is the Hausner ratio. The lower the C value, the higher the flowability.

2.1.3 Special types of core boxes

Besides the use of the PFT, also various types of test core boxes were developed to obtain more information on the flowability of sand mixtures based on inorganic binder systems (see **Figure 4**). These core boxes were manufactured aiming to measure/visualize the amount of sand mixture blown into the open space of the core box.

2.1.4 Core weight

another approach to obtain indirect data related to the flowability of sand mixtures is to measure the core weight. From a general point of view, it is suggested that the higher the core weight, the higher the compaction. Compaction, which can be given as weight per core or by density, is related to flowability. In case the flowability of a sand mixture is low, core box filling is a severe issue finally resulting in a non-complete filling of the core box resulting in low(er) density of the produced cores.

2.1.5 Core shooter

The core shooter being used during this project is a Laempe L1. This core shooter is often used to produce transverse bars but also other types of cores can be manufactured with the Laempe laboratory machine type L1 being developed for manufacturing testing cores in heated and non-heated tooling, using gas hardening processes like CO_2, cold box and hot box. The sand mixture is automatically injected into the core box, which is clamped between the side presses, and heated at various temperatures. The release of high-pressured air blows the sand from the sand storage bunker into the core box at high speed. The total elapsed shooting time for transverse bars was set at 1 s and with a shooting pressure of 4 bar. In case of producing cores, these were purged with heated air for 60 (bars) or 120 (cylinders) s and at 120°C. The

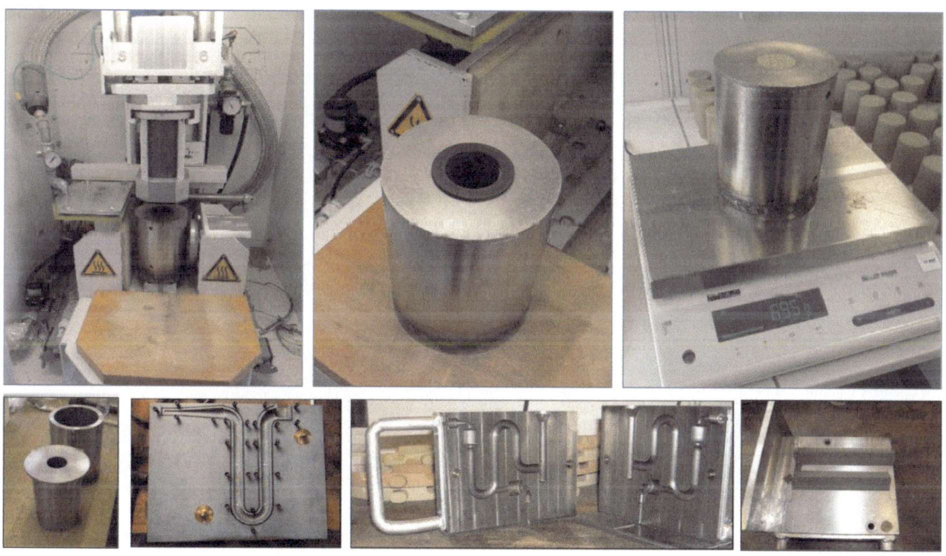

Figure 4.
Various types of core boxes with different complexities to measure flowability of different sand mixture compositions.

core box temperature was set at 140°C. However, deviations from these processing parameters are always possible since other types of cores need lower or higher shooting times and shooting pressures, shorter or longer purging times, lower or higher core box temperature, etc. In the case special types of core boxes are used to measure flowability, no curing step is considered. Pictures of this type of core blower are shown in **Figure 5**.

2.2 Various types of sand mixtures

The various types of sand mixtures can be characterized by the nature of the filling material, the type of the liquid inorganic binder and the type of additive.

2.2.1 Type of sand

In the case of sand, the most important structural parameters influencing the flowability of the sand mixtures, are the average grain size, the grain size distribution and shape. Generally, foundries will use that type of sand that is available from a local quarry near the production site, so to reduce transport costs. This means that the type of sand is a given parameter that can hardly be replaced by another type of sand. Tests presented in this contribution are mainly based on the use of H32 or H33 (Quarzwerke, Germany). These types of sand showed medium sphericity of the individual sand grains including a sub-rounded to a rounded shape. **Figure 6** shows various types of sand grains with various shapes from very angular to well-rounded and from low sphericity to high sphericity. Macrographs of quartz sand type H32 and H33 are depicted in **Figure 7**.

2.2.2 Inorganic liquid binder

Not only the type of sand, but also the type of liquid binder affects the flowability of the sand mixture. In case the viscosity of the liquid binder is relatively high, it will have a detrimental impact on flowability. Viscosity can be modified by adding solvents with specific rheological properties, or by diluting the binder easily by adding a

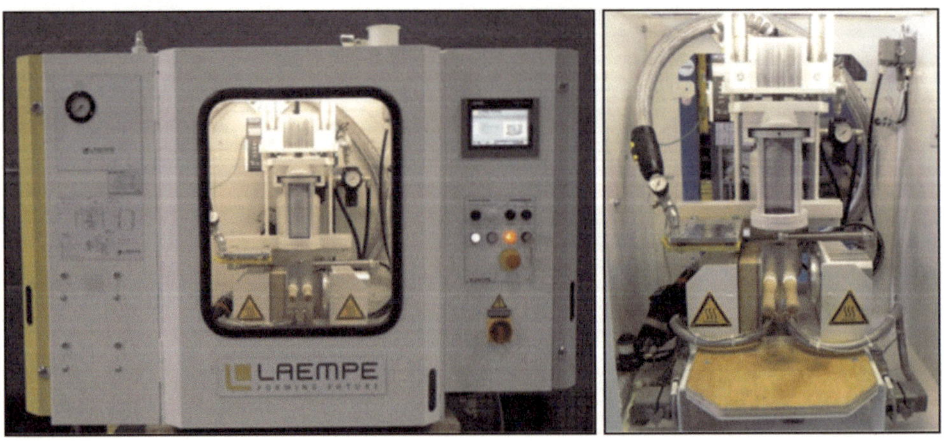

Figure 5.
Laempe L1—1-liter core blower.

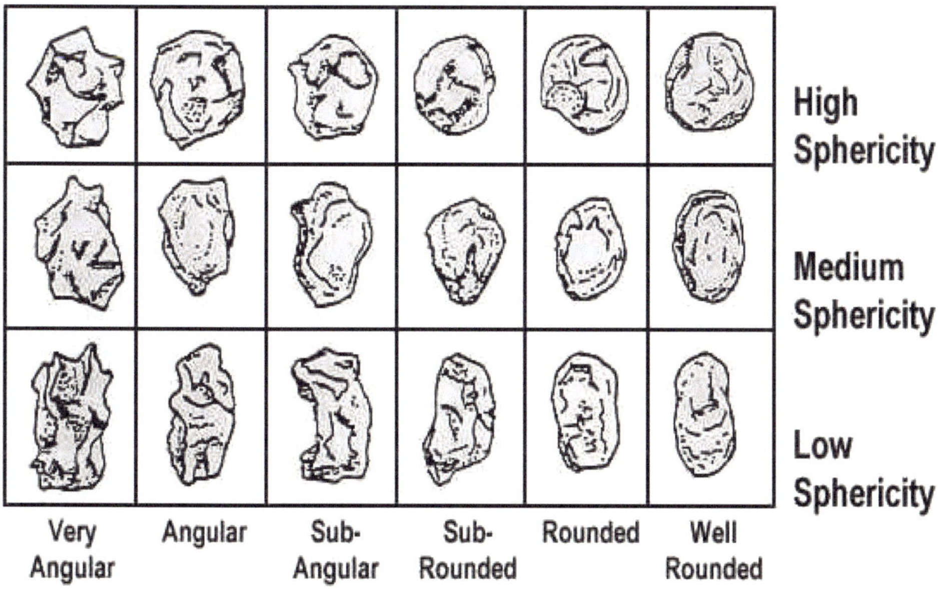

Figure 6.
Sand with various shapes from very angular to well-rounded and from low sphericity to high sphericity [5].

Figure 7.
Macrographs of sand type H32 (left) and H33 (right).

small amount of water. Adding a small amount of water of 5 wt% already lowered the kinematic viscosity by 50%.

Next to modifying or adapting viscosity, also the surface tension of the liquid can play a significant role of importance. In case the surface tension (mN/m) is rather high, it will be difficult to produce new surfaces because this is energetically unfavorable. This is the case during core manufacturing when sand is blowing into the core box. Many times, individual sand grains will be separated from each other creating new surfaces (higher state of energy). This means that it is more difficult to separate two individual sand grains from each other when surrounded by a thin liquid film of the binder with high surface energy. So, it is expected that the lower the surface energy of the binder, the better the flowability. One of the means to modify surface tension is the use of a small amount of a so-called surfactant or surface-active agent. The lowest surface tension is achieved when the CMC (critical micelle concentration) is reached, generally the addition of a few tens of percent based on the liquid binder. Once the surface area is fully occupied, the addition of more surfactant will not further reduce the surface tension of the liquid.

2.2.3 Additives to improve flowability

Special types of additives can be considered to further improve specific properties of the sand mixture as well as of the produced cores. Irregular shapes resulted generally in low flowability whereas on the other hand, spherical particles improved flowability. Often discussed in papers and patents, the use of a small amount of fly ash indeed improved flowability, which can be considered as a suitable fully inorganic mineral. Improved flowability will directly affect the bending strength due to improved compaction. Some results with various types of additives can be found in Ref. [1]. **Figure 8** shows a macro- and micrograph of a certain type of fly ash. Also appointed is the use of graphite flakes which can modify flowability, see also **Figure 9**. The presence of platelets can improve tribological properties, so these particles can also be considered as an additive of sand mixtures with inorganic binder systems. In the following, only spherical particles being used as inorganic additives will be discussed in more detail.

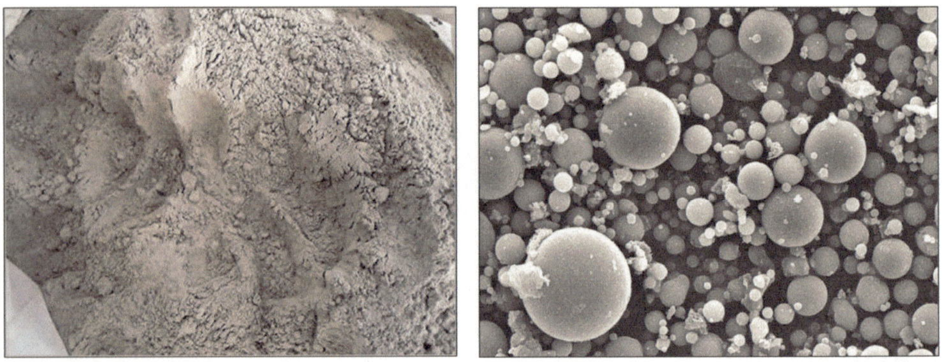

Figure 8.
Macrograph (left) and micrograph (right) of fly ash. Interesting to note is the spherical shape of the individual fly ash particles.

Figure 9.
Macrograph (left) and micrograph (right) of graphite. Interesting to note is the platelet/flaky structure of the individual graphite particles.

3. Results and discussion

3.1 On the influence of the surfactant concentration

First series of tests were done with one type of the inorganic binder and with various concentrations of three types of surfactants, A, B and C. **Figure 10** shows the surface tension as a function of the concentrations of the surfactants. Measurements with the Theta Attension optical tensiometer showed that the surface tension of the liquid inorganic binder was about 53 mN/m. The addition of 0.25 wt% bob (based on binder) surfactant type C to the liquid binder resulted already in a significant decrease of the surface tension to values of 20 mN/m. A further increase in the concentration did not result anymore in a further obvious decrease of the surface tension, since the CMC (critical micelle concentration) was already achieved. In case of surfactant type A or B, the decrease in surface tension was less pronounced.

With the use of the PFT powder flow tester, more information was obtained on the flowability properties of sand mixtures including various concentrations of the surfactant type C. Here, quartz sand type H33 was used. No additives were taken into consideration, only one type of liquid inorganic binder type. **Figure 11** shows the unconfined failure strength as a function of the applied compressive stress. This failure strength is related to the cohesive behavior of the sand mixture, and as such, it can be used to evaluate quantitative flowability. This figure shows clearly that up to about 0.2 wt% bob surfactant type C, the increase in flowability was significant. Higher concentrations showed only a further small improvement of flowability. Another approach is to calculate the Hausner ratio and the Carr index. **Table 1** shows the indices as a function of the concentration of the surfactant type C.

From **Table 1** it is clear that the Hausner, as well as the Carr index, decreased with higher concentrations of the surfactant. This relationship indicates that the flowability of the sand mixture improved with the presence of the surfactant type C, whereas with a further increase in surfactant concentration (>0.2 wt% bob) the change in these values was much less.

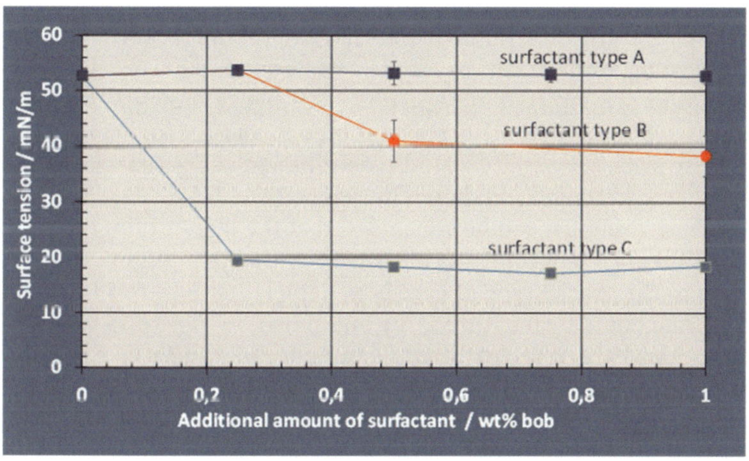

Figure 10.
Surface tension of a sodium silicate binder with various concentrations (bob: based on binder) of surfactants type A, B and C.

Conc. surf. C	0.0%	0.1%	0.2%	0.3%	0.4%	0.5%	1.0%
ρ_T	1.57	1.57	1.57	1.57	1.57	1.57	1.57
ρ_B	0.94	0.98	1.01	1.01	1.03	1.03	1.04
H	1.67	1.60	1.55	1.55	1.52	1.52	1.51
C	40.1	37.5	35.5	35.5	34.2	34.2	33.8

ρ_T = tapped bulk density (kg/dm^3); ρ_B = free density (kg/dm^3); H = Hausner ratio; C = Carr index.

Table 1.
Hausner ratio (H) and Carr index (C) as a function of the surfactant concentration (wt%—bob).

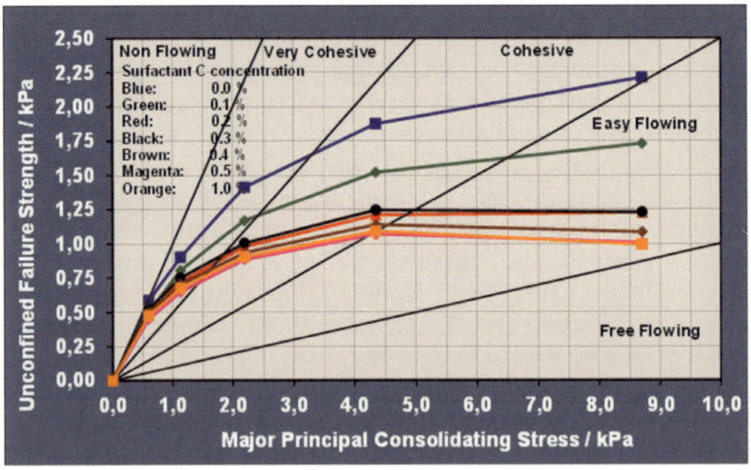

Figure 11.
Unconfined failure strength versus major principal consolidating stress as a function of the concentration of a surfactant type C.

3.2 On the influence of the type of surfactant

Not only the amount of added surfactant does influence the flowability of the sand mixture, but also the chemical composition can have an impact on this. Another series of tests were dealing with various types of surfactants. Results obtained with the PFT powder flow tester are depicted in **Figure 12**. This figure shows that the lowest flowability was achieved with the surfactant type F, and the highest with type D. In addition to these tests, flowability was also determined using a special type of core box for flowability measurements, as shown in **Figure 13**.

Based on these results, it can be concluded that the higher filling of the core box is thus related to a higher flowability as measured by the PFT (**Figure 14**).

In addition to the special type of core box, depicted in **Figure 13**, where only the weight of the sand could be measured, the one shown in **Figure 15** will also visualize the filling rate. Also here, it is clear that the presence of a surfactant resulted in higher flowability.

As already shown, the addition of a small amount of a well-chosen surfactant indeed improved the flowability of the sand mixture. This was approved and confirmed by various testing methods including powder flow tester PFT, various types of core boxes, and by calculating the Hauser ratio and Carr index. Based on these results, a series of transverse bars were manufactured with the Laempe core shooter L1.

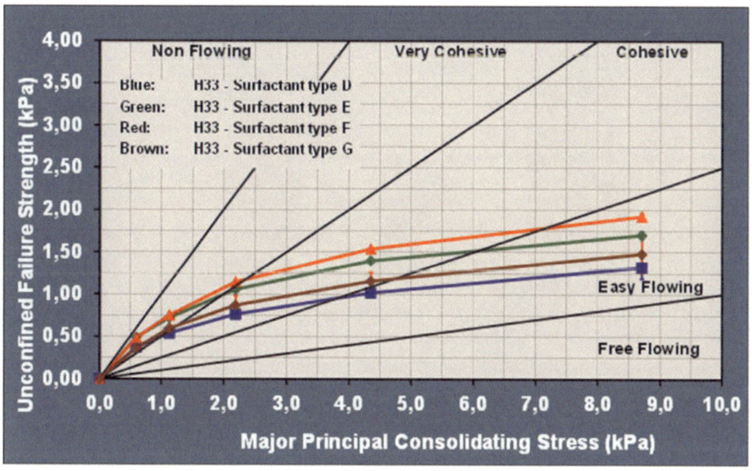

Figure 12.
Unconfined failure strength versus major principal consolidating stress as a function of the type of surfactant.

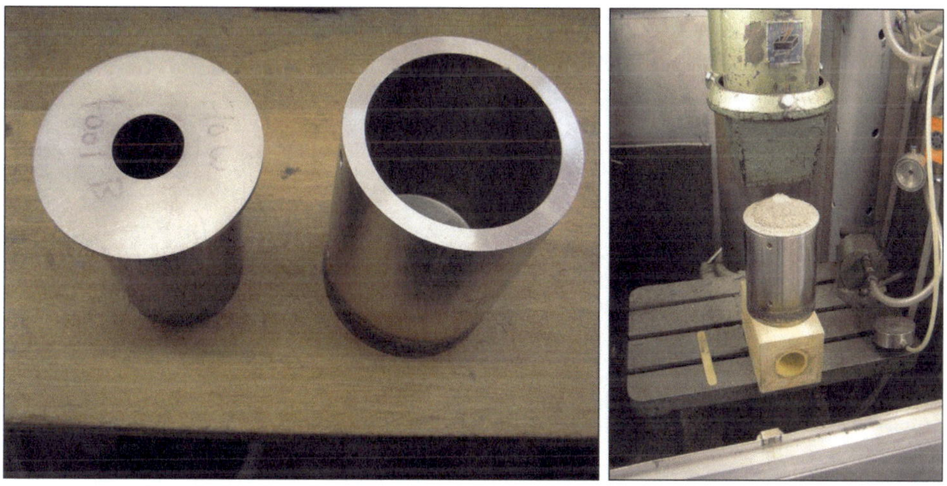

Figure 13.
Special type of core box for flowability measurements.

The outcome of these tests was that the bending strength, as shown in **Figure 16**, significantly improved with even a very low addition level of 0.05 wt% bob of the surfactant. The strength stabilized at concentration levels of 0.10 wt% bob or higher. Furthermore, also the sample weight (see also **Figure 16**) increased with the presence of a small amount of surfactant. Due to the higher compaction of the cores, the strength values were also higher, since higher compaction is directly related to a higher number of contact points (bonding bridges) between individual sand grains.

In addition to the use of surfactants, which lower the surface tension of the liquid binder and as such improve flowability, also the viscosity and the type of the liquid inorganic binder can play a significant role in the production of high-quality cores. The viscosity can be lowered by adding a small amount of water, thus reducing the solid content of the binder. Due to a lower viscosity, a more homogeneous sand mixture can be achieved. This means a more homogeneous distribution of the binder

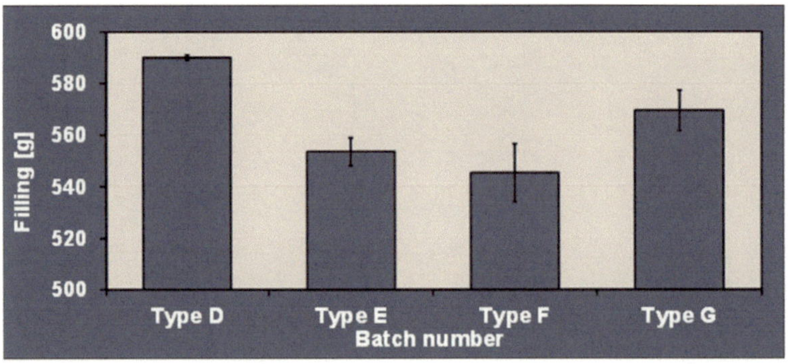

Figure 14.
Filling (weight of the sand blown into the core box) of a special type of core box as a function of the various sand mixtures.

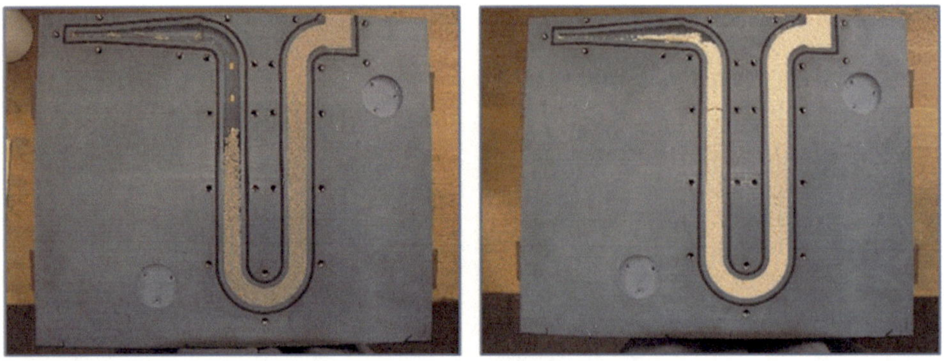

Figure 15.
An example of the filling with a sand mixture without surfactant (left) and with surfactant (right).

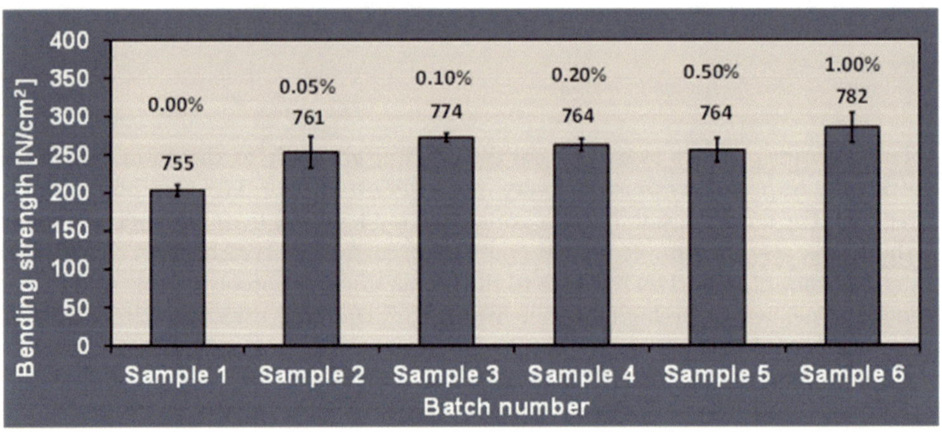

Figure 16.
Bending strength values (sample weight on top of the individual bars (in g)) as a function of the amount of surfactant added to the inorganic liquid binder (wt% bob). In this case, 1.0 wt% bos additive type A01 was also present in the sand mixture.

around the individual sand grains, which probably can result in a net lower amount of binder to achieve similar strength and surface quality. Another approach is the use of lithium—or potassium silicate, next to sodium silicate. It is well-known that the viscosity of potassium silicate is lower than that of sodium silicate, however, care must be taken since differences in strength can occur.

This means that aiming high-quality cores can be rather complex, since the various chemical and physical parameters can play a role into this. The chemical structure and the concentration of the surfactant, the viscosity and chemical composition of the liquid binder (solid content and type and modulus of the silicate) and the amount or concentration of liquid binder based on the filling material will have an impact on the flowability of the sand mixture and thus directly on core quality (compaction, surface defects, surface friability and surface smoothness).

3.3 Use of additives to improve core quality

The addition of powder-like inorganic compounds, also called core additives, to sand mixtures is at present indispensable. Since the requirements from the foundry industry are continuously growing and more demanding, sand with only a liquid inorganic binder cannot fulfill anymore all requirements. Research and development on a new generation of inorganic additives are continuing, aiming to further improve the casting process. First of all, it needs to improve the surface quality of the core, indicative of high surface smoothness. Another advantage can be improved flowability which makes cores denser. In addition to this, it will support to fill complex core boxes, such as those being used to produce complex-shaped water jackets. A well-chosen core additive can also induce higher strength values, which helps to reduce the amount of binder to be added to the sand mixture. Thus, less water is present in the sand mixture and as such, the total cycling time (read: purging time with hot air) can be further reduced. On the other hand, the casting quality is also related to the amount and type of the core additive. A non-well-chosen additive can result in amongst others, metal penetration, gas defects, low de-coring properties and sand adhesion. As already mentioned, core additives are also added to improve flowability and thus compaction and strength of the cores.

A small addition of 0.6 wt% bos of the additive type A01 resulted in an obvious improvement of flowability of the sand mixture (see **Figure 17**). However, a further increase of the concentration level resulted in over-saturation inducing lower flowability. The addition of 1.8 wt% bos type A01 (spherical particles) resulted in an even lower flowability than adding 1.2 wt% bos. A similar approach was obtained by measuring the bending strength of transverse bars including the various concentrations of the additive. These results are shown in **Figure 18**. Here, it is clear that up to about 0.3 wt% bos additive type A01 strength values increased from about 350 to 480 N/cm^2, thus an increase in strength of about 30%. Starting from 0.3 up to about 1.2 wt% bos, strength values stabilized around values between 450 and 480 N/cm^2. Higher concentrations of the additive resulted in the decrease of the bending strength.

3.4 On the influence of micro- and nano-sized spherical particles

It is well-known that the grain size (distribution) of sand being used for cores and molds can strongly influence the surface morphology of castings. Relatively large grains result in large pores allowing penetration of the liquid metal, finally resulting in

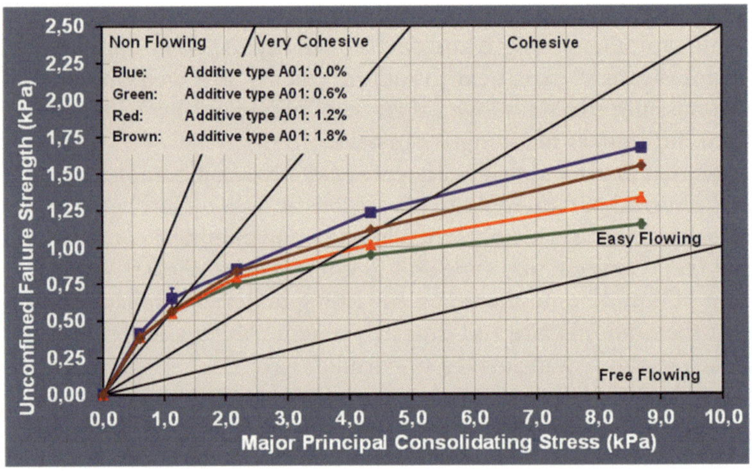

Figure 17.
Unconfined failure strength (kPa) as a function of the major principal consolidating stress (kPa) and the amount of additive type A01 (wt%—based on sand (bos)).

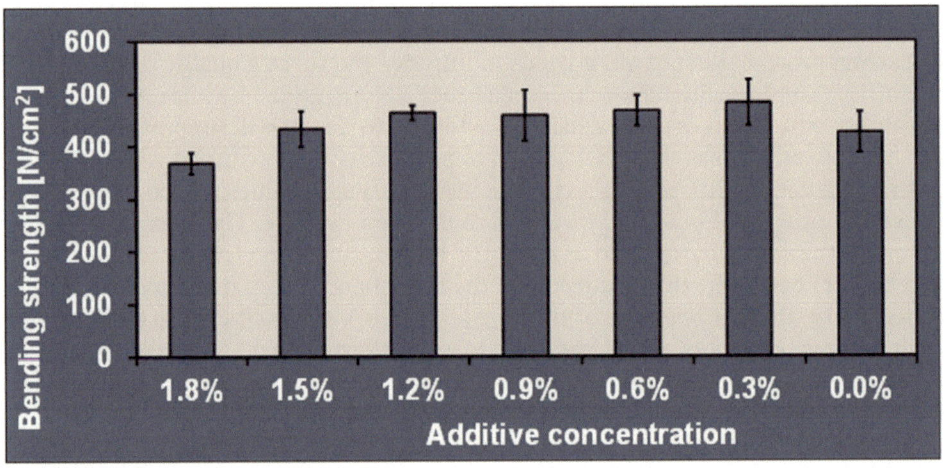

Figure 18.
Bending strength as a function of the additive concentration type A01. The liquid inorganic binder was set at 2.1 wt% (bos).

the undesired surface finish of the product. To improve the surface finish, finer grains can be chosen to suppress or inhibit metal penetration during the casting process.

However, not only the grain size distribution but also properties like the geometry and shape of the chosen sand are important. The grain shape is defined in terms of angularity and sphericity. More information concerning this can be found in the Foseco Non-Ferrous Foundryman's Handbook [6]. It is mentioned that grains with medium to high sphericity give good flowability, whereas more angular and lower sphericity of sand grains will result in low flowability, finally resulting in relatively low compactness (packing density of the sand grains).

To move individual sand grains independently of each other, relatively high friction forces Fw have to overcome. A schematic view of individual sand grains with medium to low sphericity is depicted in **Figure 19**.

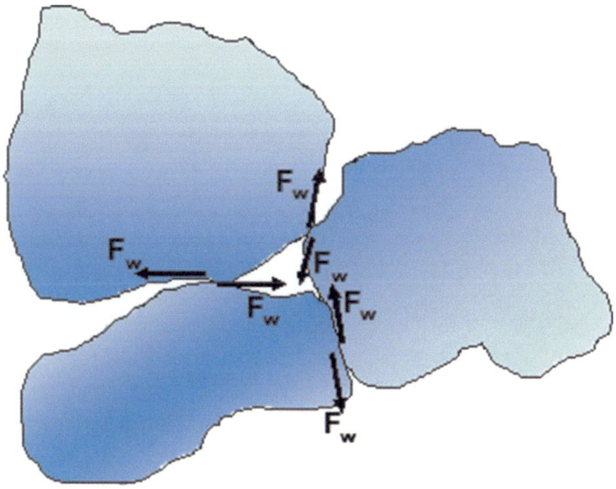

Figure 19.
Individual sand grains characterized by medium to low sphericity.

To enhance the flowability of such a type of quartz sand, one possibility is to add micro-sized spherical particles covering the outer surface of the sand grains in such a way that contact between the sand grains is obviously suppressed or even prevented. These particles will act as so-called "spacers" lowering the overall friction forces between the individual grains. In the case of inorganic additives, spherical micro-silica particles are often chosen to improve the flowability of the sand mixture. More information on flowability can be found in various references [7–12]. A micrograph of such a type of micro-silica is shown in **Figure 20**. Under the most ideal conditions, it can be expected that the relatively small particles of micro-silica cover the large sand grains by a uniform distribution.

A minimum number of these small spherical particles will be needed to achieve non-contact between the individual sand grains.

A schematic view of sand grains with the addition of micro-silica is shown in **Figure 21**. A common average diameter of quartz sand lies between 200 and 400 μm.

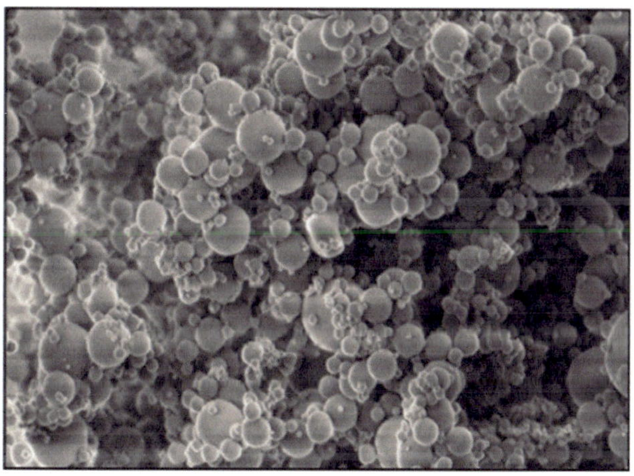

Figure 20.
Spherical micro-silica particles.

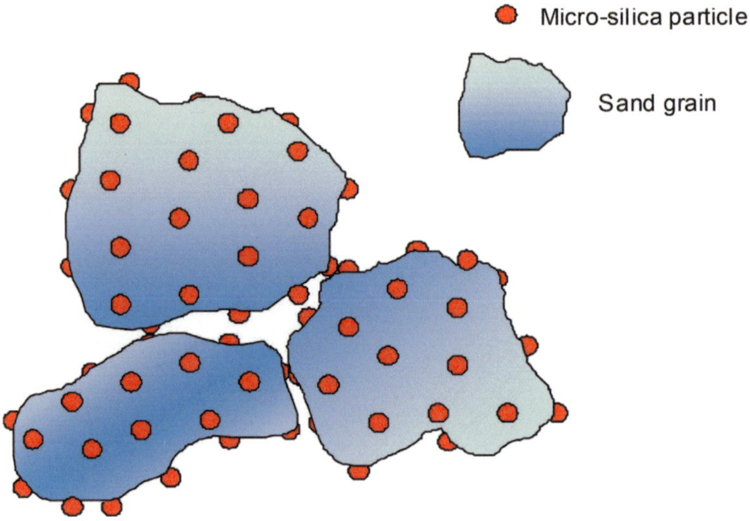

Figure 21.
Individual sand grains with the addition of micro-silica particles to ensure non-contact between the host particles.

The specific diameter of spherical particles of micro-silica is between 0.1 and 2.0 μm, significantly smaller than the sand grains. If the amount of added micro-silica is insufficient, the flowability of the sand mixture is far from the optimum, and sufficient compactness of the core and/or mold is difficult to guarantee. With the addition of a sufficient amount of micro-silica, flowability was significantly improved resulting in cores with good quality: sufficient packing density and smooth surface finish.

Based on these results, it was considered to improve further the flowability of such systems. Focusing on the guest particles, it was obvious that these were significantly smaller than the host (sand) particles. A similar model can be considered to design the mechanistic aspects of two spherical micro-silica particles. If a critical number of very small spherical particles (nano range) can be added in such a way that this cover

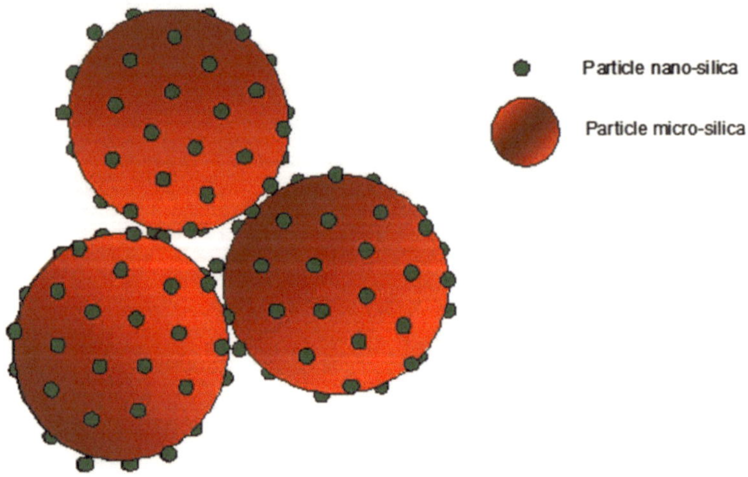

Figure 22.
Individual micro-particles (micro-silica) with the addition of nano-silica particles to ensure non-contact between the micro-silica particles.

(uniformly distributed) the outer surface, a similar configuration can be expected as that with the sand grains (see **Figure 22**). Probably due to this modified configuration, the flowability of the sand mixture systems can be improved further. However, this theory can still be discussed. A micrograph of micro-silica particles covered with nano-particles is shown in **Figure 23**. First results have indeed shown that a small increase in flowability and bending strength was achieved with the addition of a small amount of nano-particles, in this case 0.05 wt% bos.

Based on these experimental data, the following model, as depicted in **Figure 24**, can be proposed, related to an improved flowability resulting in higher compactness of the core material, and as a consequence, higher bending strength.

Besides the use of micro- and nano-spheres to improve flowability, also other compounds/components are needed to meet all requirements related to high-quality casting surfaces. It is beyond the scope of this contribution to discuss all requirements,

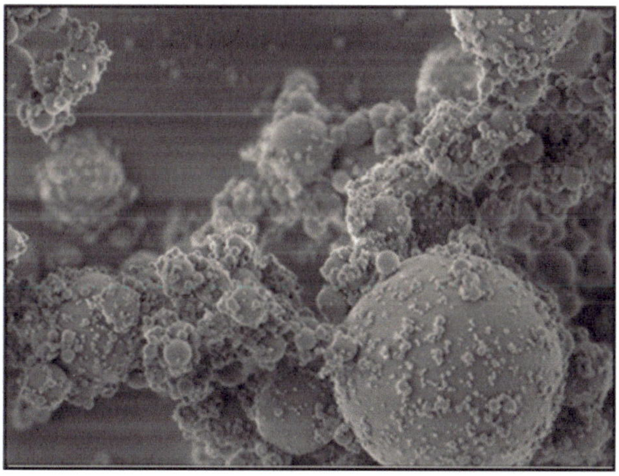

Figure 23.
Spherical micro-silica particles covered with nano-silica.

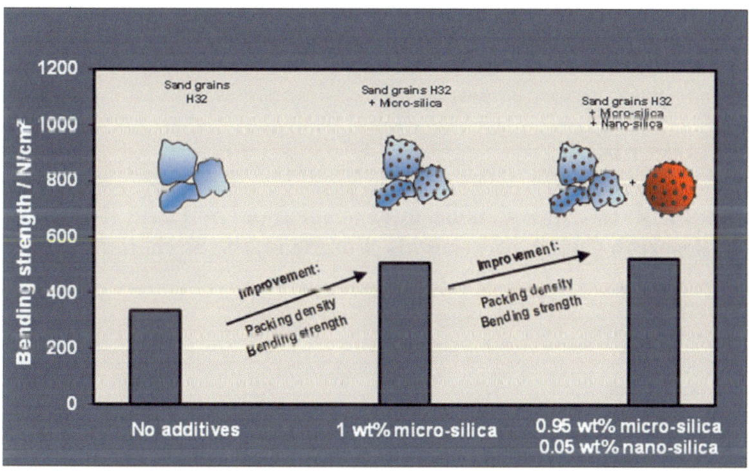

Figure 24.
Bending strength model based on the addition of small amounts of micro- and nano-silica.

but it shows that a well-balanced inorganic binder system is highly complex. In addition to high-quality cores, cold and hot strength, resistance against hot distortion, no interaction with the liquid metal, no metal penetration, good de-coring properties, etc. and finally re-use of core residue should be considered too.

4. Conclusions

Before inorganic binder systems can be considered to replace the existing organic binder systems, the combination of a liquid inorganic binder and an inorganic additive has to meet the requirements set by the foundry industry. Such a system can be extremely complex and one of the main requirements is to produce high-quality cores with sufficient compaction and with high surface smoothness. This can only be achieved when the flowability of the sand mixture is sufficient. In the case of optimizing only the flowability of the sand mixture, the liquid inorganic binder, as well as the inorganic additive, can play a significant role in this.

Results have shown that in the case of the liquid binder, various parameters can affect flowability, in particular, the presence of a small amount of a surfactant. The lower the surface tension, the higher the flowability. Also, the chemical composition of the surfactant (an-ionic, cat-ionic, or non-ionic) influences flowability.

Besides a modification of the liquid binder, also the additive can play a significant role. The size and shape of the individual particles of the additive, and also its addition rate to the sand, play also a significant role of importance. Flowability tests with the powder flow tester have clearly shown that spherical particles with specific (average) grain size with well-chosen addition rates can improve flowability finally resulting in sufficient compaction of cores and completely filling of core boxes being used for complex-shaped cores such as water jackets for cylinder heads.

The newly developed inorganic binder SOLOSIL TX by Foseco, the Netherlands, can be adapted in such a way that all requirements set by each individual foundry are met. For more complex applications, such as the production of water jackets, high flowability is needed, whereas high-quality less-complex cores can already be achieved with sand mixtures including lower flowability. In addition to flowability, also other properties of the sand mixtures and cores can be adjusted by a small modification of the environment-friendly inorganic binder system SOLOSIL TX.

Acknowledgements

The authors wish to thank all colleagues of the mold and core (M&C) team of the R&D Global Research Centre of Vesuvius, Enschede, the Netherlands, for their help and support.

Conflict of interest

The authors declare no conflict of interest.

DOI: http://dx.doi.org/10.5772/intechopen.102017

Author details

Vincent Haanappel
Foundry Global R&D Centre of Vesuvius, Foseco, Enschede, The Netherlands

*Address all correspondence to: vincent.haanappel@vesuvius.com

References

[1] Schrey A. The new environment-friendly binder system SOLOSIL TX for core making for the mass production of complex cast components. Foundry Practice, The Authoritative Magazine for Foundry Engineers. 2015;**263**:15-23

[2] Schulze D. Powders and Bulk Solids—Behavior, Characterization, Storage and Flow. Berlin, Heidelberg, Germany: Springer-Verlag; 2008. DOI: 10.1007/978-3-540-73768-1. Available from: https://www.dietmar-schulze.de/pdf/flowproperties.pdf

[3] Riley RE, Hausner HH. Effect of particle size distribution on the friction in a powder mass. International Journal of Powder Metallurgy. 1970;**6**(1):17-22

[4] Cain J. An alternative technique for determining ANSI/CEMA standard 550 flowability ratings for granular materials. Powder Handling and Processing. 2002;**14**(3):218-220

[5] Turkeli A. Sand, Sand Additives, Sand Properties, and Sand Reclamation. Available from: https://mimoza.marmara.edu.tr/~altan.turkeli/files/cpt-2-sand_sand.pdf

[6] Brown JR. Foseco Non-Ferrous Foundryman's Handbook. Linacre House, Jordan Hill, Oxford OX2 8DP, United Kingdom: Butterworth/Heinemann; 1999

[7] Xin F, Son L, Li Y. Modification of inorganic binder used for sand core-making in foundry practice. China Foundry. 2020;**17**:341-346. DOI: 10.1007/s41230-020-0018-2

[8] Zaretskiy L. Microsilica in sodium silicate bonded sands. International Journal of Metalcasting. 2019;**13**:58-73. DOI: 10.1007/s40962-018-0247-y

[9] Ng Y, Ng P, Wong H, Kwan A. Roles of silica fume and fly ash in improving flowability, segregation stability and passing ability of self-consolidating concrete. In: Proceedings of the 9th CANMET/ACI International Conference on Fly Ash, Silica Fume, Slag and Natural Pozzolans in Concrete; Warsaw, Poland; 20–25 May 2007. Michigan: American Concrete Institute; 2007. pp. 109-123

[10] Chang Y, Hocheng H. The flowability of bentonite bonded green molding sand. Journal of Materials Processing Technology. 2001;**113**: 238-244

[11] Ni C, Lu G, Jing T, Wu J. Influence of core sand properties on flow dynamics of core shooting process based on experiment and multiphase simulation. Research and Development, China Foundry. 2017;**14**(2):121-127. DOI: 10.1007/s41230-017-6118-y

[12] Anwar N, Sappinen T, Jalava K, Orkas J. Comparative experimental study of sand and binder for flowability and casting mold quality. Advanced Powder Technology. 2021;**32**(6):1902-1910. DOI: 10.1016/j.apt.2021.03.040

Chapter 4

Novel Physical Modelling under Multiple Dimensionless Numbers Similitudes for Precise Representation of Molten Metal Flow

Yuichi Tsukaguchi, Kodai Fujita, Hideki Murakami and Roderick I.L. Guthrie

Abstract

Physical model experiments, together with numerical model calculations, are essential for scientific investigations such as molten metal flow in casting processes. Considering the physical modelling of flow phenomena, a common method is used to construct a physical model with a reduced scale ratio and then, experiment is carried out under one or two dimensionless number(s) similitude(s). It is an ideal condition of the experiment to establish the simultaneous similitude of multiple dimensionless numbers (SMDN) concerned with the objective flow phenomena but was considered difficult or impossible to realize in practice. This chapter presents a breakthrough in this matter. A simple relationship between the physical properties of fluids and the scale ratio of the physical model is clearly expressed for the simultaneous similitude of the Froude, Reynolds, Weber, Galilei, capillary, Eötvös and Morton numbers. For establishing the physical modelling to represent molten Fe flow phenomena under the SMDN condition, the physical properties of some molten metals can be demonstrated to meet the required relationships. Furthermore, this novel concept is also applicable for other combinations of molten metals. Precise, safe, and easy physical model experiments will be conducted under the SMDN condition that exactly mimics industrial casting operations in higher-temperature systems.

Keywords: physical modelling, physical model experiment, similitude, similarity, viscosity, surface tension, density, Reynolds number, froude number, weber number, flow phenomena in mould

1. Introduction

Molten metal flow in casting mould has great influence on the productivity and quality of the cast products. **Figure 1** shows an image of flow phenomena in a continuous casting (CC) mould of steel production. There are various flow phenomena described in **Figure 1**; closed channel flow contains small vortex in submerged entry nozzle (SEN), injected argon (Ar) behaviour, free-surface flow, viscous bulk flow, and so on. It is difficult to observe or measure the velocity of high-temperature opaque molten metal flow. Consequently, physical model experiments, as well as numerical simulations, are widely carried out to estimate the flow phenomena in the mould.

As for physical modelling, similitude is an essential matter for accurate representation. Prior dimensionless numbers to represent flow phenomena in the continuous casting mould are Froude number (Fr), Reynolds number (Re), and Weber number (We) [1–4]. However, the simultaneous similitude of these three dimensionless numbers has been recognised as being difficult or impossible to realise [5–7]. As a matter of record, there has been no concept previously proposed for similitudes of multiple dimensionless numbers in physical modelling [2–36].

Standing on this point of view, physical modelling condition of simultaneous similitude of multiple dimensionless numbers (SMDN) has been studied and then

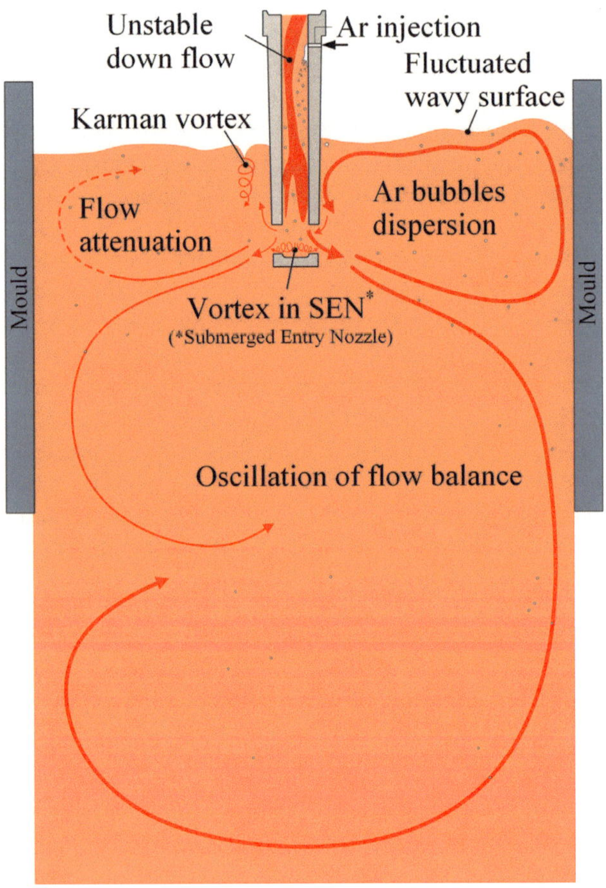

Figure 1.
Image of molten steel flow in CC mould.

established [37, 38] for the precise representation of the real flow phenomena with an appropriate relationship of scale ratio and physical properties of liquids.

2. Physical modelling condition to satisfy simultaneous similitude of multiple dimensionless numbers

Bulk flow is governed by inertial, gravitational, and viscous forces. Consequently, dimensionless numbers must correspond with are the Fr and the Re to represent the bulk flow. In addition, the We is important for the precise representation of the surface flow phenomena, such as droplet/bubble formations or ripple waves.

These three dimensionless numbers are defined in Eqs. (1), (2), and (3) [39], where the Fr is defined as the square root of (inertial force/gravitational force), the Re is defined as (inertial force/viscous force), and the We is defined as the square root of (inertial force/surface-tension force). Alternative definitions for the Fr (inertial force/ gravitational force) and We (inertial force/surface-tension force) attain the same result.

$$\mathrm{Fr} = \frac{V}{\sqrt{g \bullet L}} \tag{1}$$

$$\mathrm{Re} = \frac{\rho \bullet V \bullet L}{\eta} \tag{2}$$

$$\mathrm{We} = V\sqrt{\frac{\rho \bullet L}{\sigma}} \tag{3}$$

where V is characteristic velocity of fluid, L is characteristic length, g is gravitational acceleration, ρ is density of fluid, η is viscosity of fluid, and σ is surface tension of fluid.

Describing the prototype with subscript 0, flow velocity V in the physical model is expressed as Eq. (4) in the case of the Fr similitude ($\because Fr = \frac{V_0}{\sqrt{g \bullet L_0}} = \frac{V}{\sqrt{g \bullet L}} = \frac{V}{\sqrt{g \bullet \lambda \bullet L_0}}$) [8].

$$V = \sqrt{\lambda} \bullet V_0 \tag{4}$$

where λ is scale ratio of the physical model.

Coincidence of the Re between a physical model and the prototype are expressed as Eq. (5).

$$\mathrm{Re} = \frac{\rho_0 \bullet V_0 \bullet L_0}{\eta_0} = \frac{\rho \bullet V \bullet L}{\eta} = \frac{\rho \bullet V \bullet \lambda \bullet L_0}{\eta} \tag{5}$$

In the same manner, coincidence of the We between a physical model and the prototype is expressed as Eq. (6).

$$\mathrm{We} = \mathrm{V}_0\sqrt{\frac{\rho_0 \bullet L_0}{\sigma_0}} = \mathrm{V}\sqrt{\frac{\rho \bullet L}{\sigma}} = \mathrm{V}\sqrt{\frac{\rho \bullet \lambda \bullet L_0}{\sigma}} \tag{6}$$

Scale ratio λ for the Fr and Re similitudes obtained by substituting Eq. (4) into Eq. (5) and scale ratio λ for the Fr and We similitudes obtained by substituting Eq. (4)

into Eq. (6) should be equal in the condition of the Fr, Re, and We similitudes [37, 38]. Subsequently, Eq. (7) is obtained to express a relationship of physical properties for the simultaneous similitude of multiple dimensionless numbers (SMDN).

$$\lambda = \left(\frac{\eta/\rho}{\eta_0/\rho_0}\right)^{\frac{2}{3}} = \left(\frac{\sigma/\rho}{\sigma_0/\rho_0}\right)^{\frac{1}{2}} \tag{7}$$

Eq. (7) can be rearranged to Eq. (8) by applying kinematic viscosity $\nu = \eta/\rho$ and a new parameter of $\iota = \sigma/\rho$. $\iota^{1/2}/\nu^{2/3}$ in Eq. (8) is a parameter (named the S-parameter) to indicate satisfaction of the SMDN condition [37]. Precise physical model experiments will be carried out employing fluids that satisfy the relationship of physical properties described in Eq. (8) with the appropriate scale ratio shown in Eq. (7).

$$\frac{\iota^{1/2}}{\nu^{2/3}} = \frac{\iota_0{}^{1/2}}{\nu_0{}^{2/3}} \tag{8}$$

As a result of the Fr, Re, and We similitudes satisfying Eqs. (7) and (8), the similitude conditions of Eötvös number (Eö) and the Morton number (Mo) are also established as described below.

As shown in Clift, Grace, and Weber's chart that classifies the shape of the rising bubble [40], the Eö and Mo have a dominant influence on the bubble shape and the bubble rising velocity.

The Eö described in Eq. (9) [40], represents buoyancy force/surface-tension force. If the density of the gas is sufficiently smaller than that of the surrounding liquid, Eq. (9) can be converted to Eq. (10). Where subscript b is for the gas phase of the bubble.

The rearranged right side of Eq. (10) as shown in Eq. (11) indicates that the Eö similitude is established under the condition of the Fr and We similitudes [37, 38]. Here, the value of d_b/L in Eq. (10) is same for the prototype and the physical model ($\because d_b/L = \lambda d_{b0}/\lambda L_0 = d_{b0}/L_0$).

$$\text{Eö} = \frac{g(\rho - \rho_b) \bullet d_b{}^2}{\sigma} \tag{9}$$

$$\text{Eö} \approx \frac{g \bullet \rho \bullet d_b{}^2}{\sigma} \tag{10}$$

$$\frac{g \bullet \rho \bullet d_b{}^2}{\sigma} = \frac{V^2 \bullet \rho \bullet L}{\sigma} \bullet \frac{g \bullet L}{V^2} \bullet \left(\frac{d_b}{L}\right)^2 = \left(\frac{\text{We}}{\text{Fr}} \bullet \frac{d_b}{L}\right)^2 \tag{11}$$

Provided that the density of the bubble is quite little compared with that of the surrounding fluid, Mo described in Eq. (12) [1, 41] consists of gravitational acceleration g and the physical properties of the fluid. Accordingly, the Mo similitude is undoubtedly established in the case that the physical model employs the same fluid as in the prototype. In contrast, the physical properties should satisfy a strict relationship for the Mo similitude in the case that the fluids are different for the physical model and the prototype. As described in Eq. (13), Mo is comprised of three dimensionless numbers; Fr, Re, and We [1, 41]. As such, Mo similitude is always established under the SMDN condition for the case of simultaneous dimensionless numbers equalities for the Fr, Re, and We.

$$\text{Mo} = \frac{g \bullet \eta^4}{\rho \bullet \sigma^3} \tag{12}$$

$$\frac{\mathrm{We}^6}{\mathrm{Re}^4 \bullet \mathrm{Fr}^2} = \frac{V^6 \bullet \rho^3 \bullet L^3}{\sigma^3} \bullet \frac{\eta^4}{\rho^4 \bullet V^4 \bullet L^4} \bullet \frac{g \bullet L}{V^2} = \mathrm{Mo} \tag{13}$$

Similarly, any other combinations of inertial, gravitational, viscous, and surface tension forces, such as gravitational/viscous forces (Galilei number) and viscous/surface tension forces (the capillary number), are identical for the physical model and the prototype, under the above mentioned SMDN condition.

3. Physical properties of molten metals and water

Precise physical model experiment to represent flow phenomena under the SMDN (similitude of multiple dimensionless numbers) condition could be realised in the case that physical properties of molten metal in the prototype and experimental fluid satisfy the strict relationship described in Eq. (8).

Table 1 shows the physical properties of molten metals and water reported by many researchers and scientists [42–54]. Where "Sn-40Bi" in **Table 1** means 60mass% Sn – 40mass% Bi alloy.

These physical properties listed in **Table 1** are plotted in **Figure 2**, where the abscissa X (horizontal)-axis and the ordinate Y (vertical)-axis are the denominator and numerator of the S-parameter ($\iota^{1/2}/\nu^{2/3}$), respectively. This $\nu^{2/3}$ versus $\iota^{1/2}$ chart has been named the SMDN chart. The SMDN chart indicates whether a combination of two liquids satisfies the relation in Eq. (8), or not. As indicated in **Figure 2**, Fe at 1575°C, Cu (1085–1185°C, closely at 1145°C), and Sn (232–332°C, closely at 288°C) are all located on the same S-parameter line of $\iota^{1/2}/\nu^{2/3} = 211$. This indicates that the SMDN condition is established between these three molten metals at the appropriate temperatures and scale ratios.

	Melting point [°C]	Temperature [°C]	ρ [kg/m³]	η [Pa·s]	σ [N/m²]
			×10³	×10⁻³	
Fe	1538 [42]	1575	7.001 [42]	4.732 [43]	1.844 [43]
Hg	−38.9 [44]	20	13.55 [44]	1.568 [44]	0.4802 [45]
Sn	232 [45]	232 ~ 332	6.971 ~ 6.907 [46]	2.302 ~ 1.713 [46]	0.5645 ~ 0.5527 [46]
Cu	1085 [47]	1085 ~ 1185	7.997 ~ 7.915 [47]	4.019 ~ 3.417 [47]	1.2576 ~ 1.237 [48]
Zn	420 [49]	500	6.488 [49]	3.072 [49]	0.7507 [50]
Al	660 [51]	700	2.379 [52]	1.295 [51]	0.8694 [53]
Ga	29.8 [49]	70	6.053 [49]	1.407 [49]	0.7140 [45]
Bi	271 [54]	320	9.986 [54]	1.644 [54]	0.3758 [54]
Sn-40Bi	174 [54]	225	7.900 [54]	2.219 [54]	0.4551 [54]
Water	0	10 ~ 70	1.000 ~ 0.978 [44]	1.307 ~ 0.404 [44]	0.0742 ~ 0.0644 [44]

Table 1.
*Physical properties of molten metals and water for **Figure 2**.*

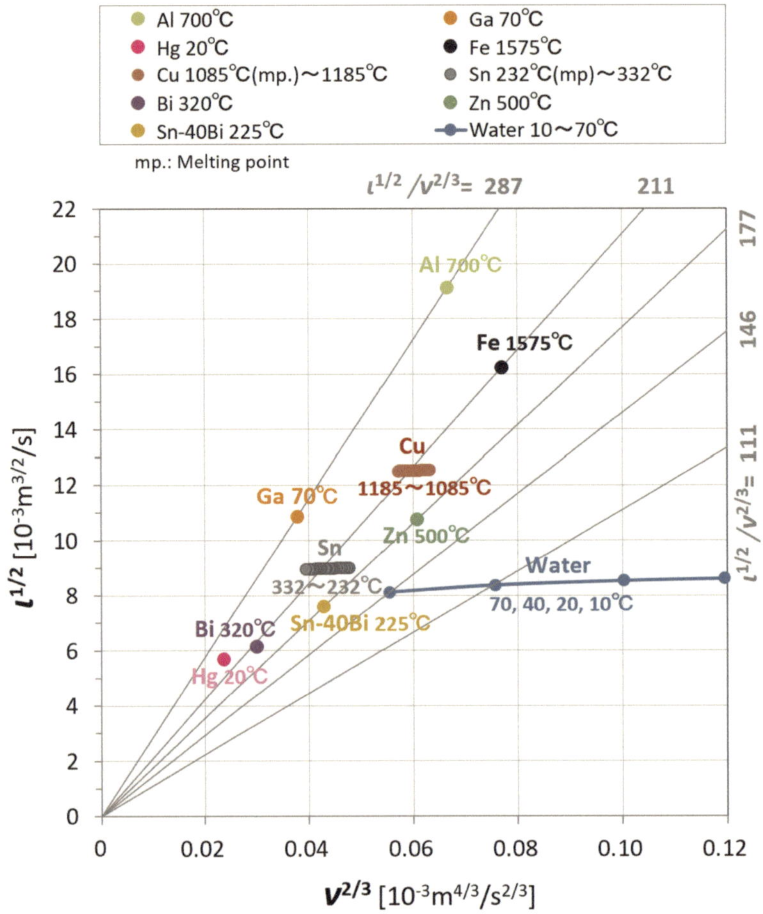

Figure 2.
Physical properties of molten metals and water plotted on SMDN chart.

Scale ratios to satisfy the SMDN conditions are also indicated in the SMDN chart. For example, the ratio of the distance between the origin point and the Ga plot to the distance between the origin point and the Al plot is 0.57, as shown in **Figure 3**. As is obvious from Eq. (7), this value of 0.57 is the scale ratio of molten Ga models needed to represent molten Al flow satisfying the SMDN condition. In the same manner, molten Sn-40Bi models with a scale ratio of 0.71 can represent molten Zn flow under the SMDN condition. Naturally, the same rule is applicable for molten Fe, Cu, and Sn.

In the case only the Fr and Re similitudes are required to represent the bulk flow, water models could represent flow phenomena of molten Fe, Cu, Sn, Al, Ga, Zn, Sn-40Bi, Bi, and Hg with proper scale ratios according to the ratios of $\nu^{2/3} = (\eta/\rho)^{2/3}$, as shown in Eq. (7). Please note that kinematic-viscosity ν of water has a large temperature-dependency, as shown in **Figure 2**. Therefore, temperature of the water should be carefully considered to determine the scale ratio.

Moreover, the physical modelling concept of the SMDN is applicable not only for molten metal systems described in this chapter but also for many types of liquids [37].

4. Applications of physical modelling under multiple dimensionless numbers similitudes

As described above, the physical modelling concept of SMDN is applicable for flow phenomena influenced by followed four forces—inertial, gravitational, viscous, and surface tension forces. Some examples of flow phenomena representations under the SMDN condition are shown in this section.

4.1. Injected bubble size

In the case of bubble formation during gentle gas injection, three forces—gravity, surface tension, and buoyancy forces, are dominant to determine the bubble diameter D_b, and thus D_b can be calculated using Eq. (14) [55], provided that the density of the gas bubble is negligibly small compared with that of the surrounding fluid.

$$D_b = \sqrt[3]{\frac{6 \bullet D_n \bullet \sigma}{\rho \bullet g}} \tag{14}$$

where D_n is the internal diameter of the gas injection nozzle.

Table 2 shows the parameters of physical properties (ν and ι), the S parameters described in Eq. (8), and the adequate scale ratios obtained by Eq. (7) for several molten metals. The physical properties of these molten metals are shown in **Table 1**. As is obvious in **Table 2**, there are three groups with similar S-parameters—molten Al and Ga for the S-parameter of 287, molten Fe, Cu, and Sn for the S-parameter of 211, and molten Zn and Sn-40Bi for the S-parameter of 177. Scale ratios shown in **Table 2** are defined against "BASE" molten metals for each S-parameter group. The "BASE" metals are—molten Al for the group of S-parameter = 287, molten Fe for the group of S-parameter = 211, and molten Zn for the group of S-parameter = 177.

Bubble diameters formed by injected inert gas under those conditions calculated by Eq. (14) for various nozzle diameters are shown in **Figure 4**. The abscissa X (horizontal)-axis of **Figure 4** shows equivalent bubble diameters considering the scale ratios. As is clear from **Figure 4**, equivalent bubble diameters are plotted on the same point under the conditions with the same S-parameters. This is a simple comparative case to show the availability of physical model experiments respecting the SMDN condition.

4.2. Rising velocity of bubbles

Inert gas injection is widely applied to casting processes to remove non-metallic inclusions. Meanwhile, it causes bubble defects of the cast products. Consequently, it is valuable to represent the two-phase flow of bubbles and molten metals in physical model experiments.

Complicated demeanour of the rising bubble is dominated by two dimensionless numbers—the Eö and Mo. They have dominant influences on the bubble shape as well as the bubble rising velocity [40]. Therefore, physical models under the concept of SMDN will accurately represent bubble rising demeanour.

In connection with the bubble deformation, the drag coefficient of a rising bubble changes in value complicatedly, which has been investigated by many researchers. Eq. (15) shows Tomiyama's Equation [56], one of the trustworthy equations to express

	Temperature [°C]	ν [m²/s]	ι [m³/s²]	S-parameter = $\iota^{1/2}/\nu^{2/3}$	Scale ratio $\lambda = (\nu/\nu_0)^{2/3}$
		$\times 10^{-7}$	$\times 10^{-4}$		
Al	700	5.441	3.655	287	1.0 (BASE)
Ga	70	2.325	1.180	287	0.57
Fe	1575	6.759	2.635	211	1.0 (BASE)
Cu	1145	4.716	1.568	211	0.77
Sn	288	2.776	8.045	211	0.55
Zn	500	4.734	1.157	177	1.0 (BASE)
Sn-40Bi	225	2.809	5.761	177	0.71

Table 2.
Physical properties and S-parameters for various molten metals.

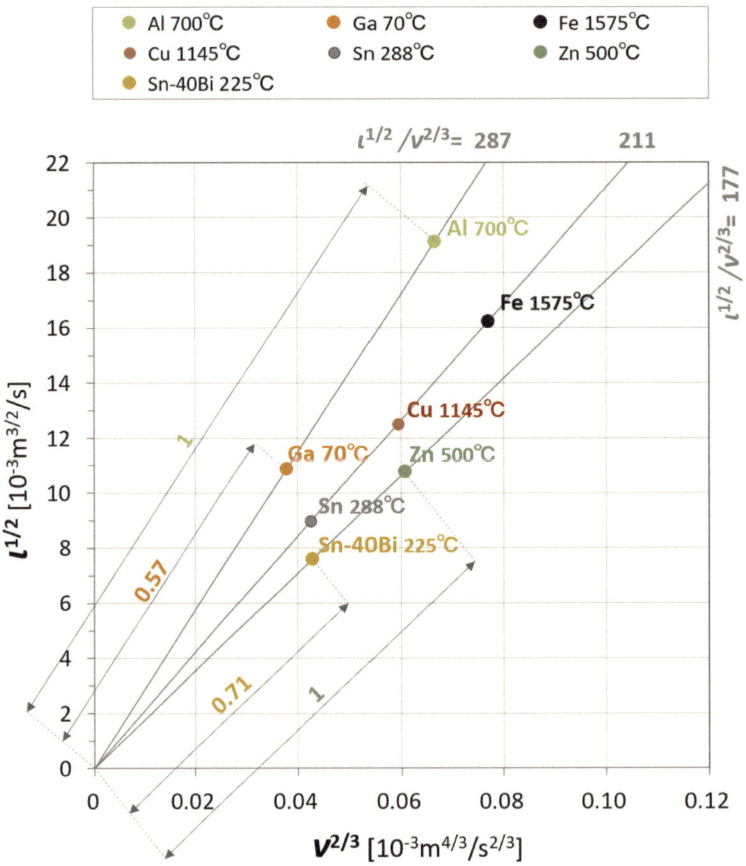

Figure 3.
Proper scale ratios indicated on SMDN chart.

the drag coefficient of rising bubbles. However, this equation was obtained by the experiments with water and some organic liquids, so it should be confirmed if this equation could be applied to the molten metal systems or not.

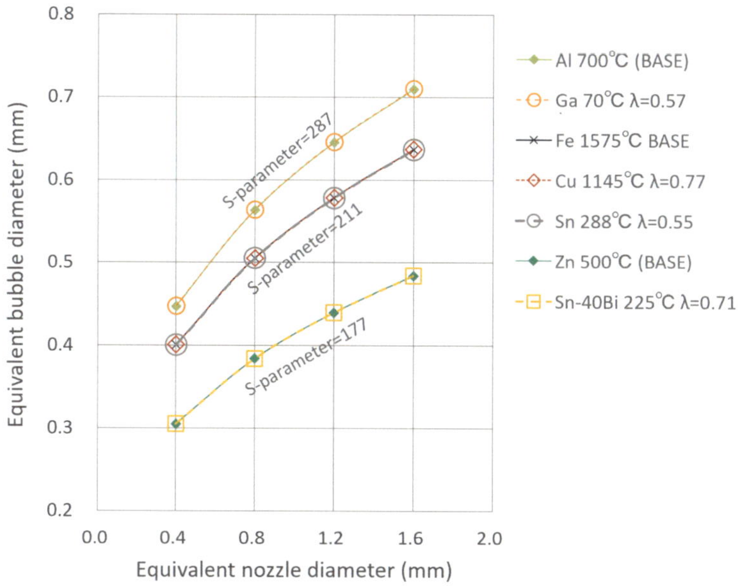

Figure 4.
Generated bubble sizes for various S-parameter groups.

$$C_D = \max\left[min\left\{\frac{24}{\mathrm{Re}_b}\left(1 + 0.15\,\mathrm{Re}_b{}^{0.685}\right), \frac{72}{\mathrm{Re}_b}\right\}, \frac{8}{3}\frac{\text{Eö}}{\text{Eö} + 4}\right] \tag{15}$$

Therefore, data of investigated terminal velocity of rising bubble in molten metals of reliable researches [38, 57] were compared with Tomiyama's equation. As shown in **Figure 5**, Tomiyama's equation well represents the bubble terminal velocity in molten Sn and Hg. The result shows that Tomiyama's equation can be extensively applied to molten metal systems.

Accordingly, bubble rising velocity in molten metals could be calculated by Tomiyama's equation. Before comparing the bubble rising velocity, it should be noted that in the case of the Fr similitude with the scale ratio of physical modelling, flow velocity is described in Eq. (4; $V = \sqrt{\lambda} \bullet V_0$). Consequently, time progression in the physical model will expand or contract as expressed in Eq. (16) [38]. For example, in the case of scale ratio $\lambda = 0.5$, time progression in the physical model will shrink $\sqrt{0.5} \approx 0.71$ times as compared with the real-time progression. For example, in the case of the physical models under the Fr similitude and the scale ratio of 0.5, real flow phenomena generated in 100 seconds will be represented in 71 seconds.

$$t = \frac{L}{V} = \frac{\lambda \bullet L_0}{V_0\sqrt{\lambda}} = \sqrt{\lambda} \bullet t_0 \tag{16}$$

Figure 6 shows the relationship between equivalent bubble diameter in molten Fe and relative rising velocity (*RRV*) of bubbles under variable similitude conditions. *RRV* is (rising velocity of the bubble in physical model experiment considering the scale ratio/rising velocity of the bubble in prototype) as expressed in Eq. (17). The time progression described in Eq. (16) is also taken into account in Eq. (17). Here, U_0 is bubble rising velocity in the prototype (in molten Fe), U is bubble rising velocity in

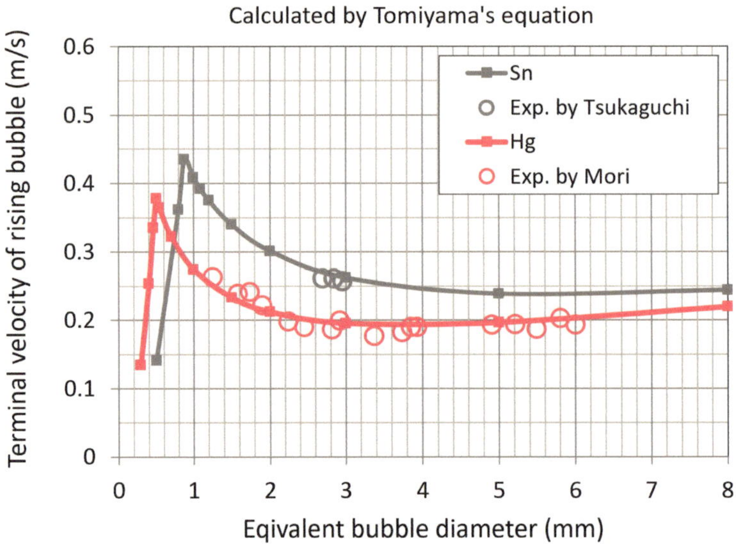

Figure 5.
Comparison of terminal velocity of rising bubbles in molten metals.

the physical model experiment. Please note that the bubble diameter d_b for U and bubble diameter d_{b0} for U_0 are not same ($d_b = \lambda \bullet d_{b0}$).

$$RRV = \left({}^{\frac{U}{\lambda}}\!/_{U_0} \right) \sqrt{\lambda} = \frac{U}{U_0 \bullet \sqrt{\lambda}} \tag{17}$$

As shown in **Figure 6**, rising velocity of bubbles in molten Sn or molten Cu under the SMDN condition precisely represents rising velocity of bubbles in molten Fe. On the other hand, rising velocity of bubbles in water under the Fr and Re similitudes condition ($\lambda = 1.0$) corresponds to that in molten Fe only in the region of the small bubble diameters. In contrast, rising velocity of bubbles in water under the Fr and We similitudes condition ($\lambda = 0.52$) correspond to that in molten Fe only in the region of the large bubble diameters. The reason can be read from Eq. (15). Drag coefficient of bubbles are dominated by the Re in the region of the small bubble diameters, consequently, the water model under the Fr and Re similitudes condition is enough to represent the small-bubble rising velocity. In the region of the large bubble diameters, drag coefficient of bubbles is dominated by the Eö, and the Eö similitude is established under the Fr and We similitudes as described in Eqs. (10) and (11). Consequently, the water model under the Fr and We similitudes condition is adequate to represent the large-bubble rising velocity.

4.3. Other applications of physical modelling under multiple dimensionless numbers similitudes

Other applications of the SMDN concept include the following examples—in the case of the Mesler entrainment of a liquid drop to produce many fine bubbles, three-dimensionless numbers, the We, Fr, and capillary numbers, describe the entrainment phenomenon [58]. Consequently, physical model experiments under the SMDN condition will precisely represent the bubble sizes formed by the Mesler entrainment.

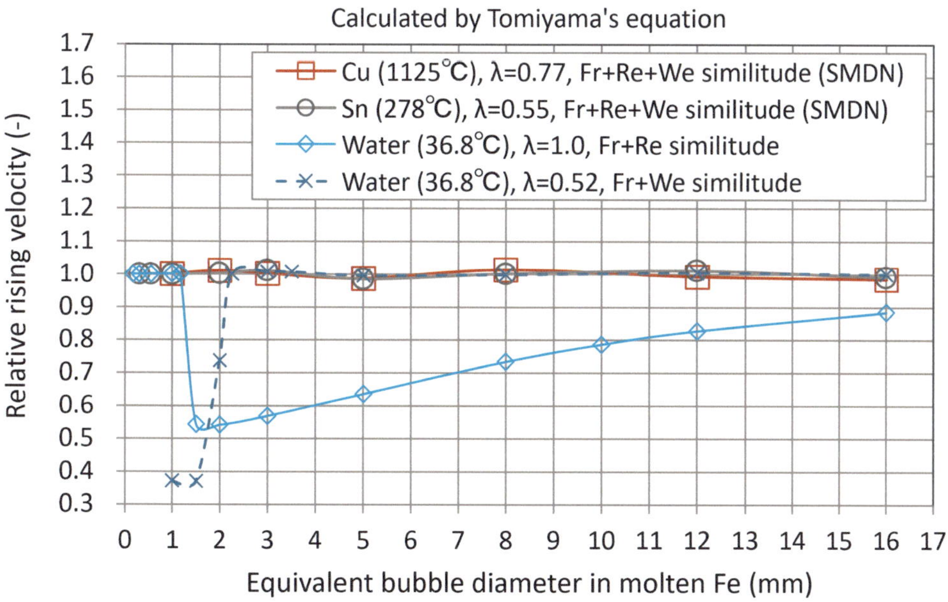

Figure 6.
Comparison of relative terminal velocity of rising bubbles under a variety of similitude conditions.

As expressed in the above examples, the SMDN concept of the physical modelling is applicable for the precise representation of all the flow phenomena dominantly affected by four forces—inertial force, gravitational force, viscous force, and surface-tension force. However, physical models under the SMDN condition could not precisely represent flow phenomena affected by other forces or flow phenomena of other types of fluids. For example, flow phenomena of compressible fluid flow, two-phase flow with solid-particles and liquid are outside the SMDN concept region in this chapter.

As for the representation of the thermal convection flow, the similitude of the Rayleigh number (Ra), shown in Eq. (18) for both the prototype and the physical model [59], is required.

$$\mathrm{Ra} = \frac{g \bullet \beta_0 \bullet \Delta T_0 \bullet {L_0}^3}{\nu_0 \bullet \alpha_0} = \frac{g \bullet \beta \bullet \Delta T \bullet L^3}{\nu \bullet \alpha} \tag{18}$$

where α is thermal diffusivity, β is thermal expansion coefficient, and ΔT is temperature drop across the convection layer.

The thermal convection flow phenomena will also be represented under the SMDM condition in the case that ΔT was controlled to make the Rayleigh numbers equal expressed in Eq. (18).

Eq. (18) can be rearranged to Eq. (19) by substituting $L = \lambda \bullet L_0$, Eq. (7) that expresses the scale ratio of the SMDN condition and kinematic viscosity $\nu = \eta/\rho$. Then theΔT ratio to satisfy the Rayleigh number similitude under the SMDN condition is obtained, as shown in Eq. (20) [37]. It should be noted that if the ΔT ratio calculated by Eq. (20) was large, thermal effects on the physical properties of the experimental liquid could not be negligible. If temperature dependencies of the physical properties of two liquids for the prototype and the physical model are similar, a large ΔT_0 itself will not induce a lower accuracy in the experimental results.

$$\text{Ra} = \frac{g \bullet \beta_0 \bullet \Delta T_0 \bullet L_0{}^3}{\nu_0 \bullet \alpha_0} = \frac{g \bullet \beta \bullet \Delta T \bullet (\lambda \bullet L_0)^3}{\nu \bullet \alpha} = \frac{g \bullet \beta \bullet \Delta T \bullet \left(\nu/_{\nu_0}\right)^2 \bullet L_0{}^3}{\nu \bullet \alpha} \quad (19)$$

$$\frac{\Delta T}{\Delta T_0} = \frac{\alpha \bullet \beta_0 \bullet \nu_0}{\alpha_0 \bullet \beta \bullet \nu} \quad (20)$$

Naturally, the SMDN concept of physical modelling is applicable to represent the turbulent flow. Minutes of turbulent flow, such as small eddies, vortexes, and swirls, will be precisely represented since such flow phenomena are governed by inertial, gravitational, and viscous forces. It should be noted that the thermal effects on the physical properties should be considered under heavy turbulent flows with the rising temperature of the experimental fluid. Besides, the wettability between the fluid and the wall of the flow channel will affect the result.

5. Conclusion

The novel physical modelling concept, involving the simultaneous similitude of multiple dimensionless numbers, the Froude, Reynolds, Weber, Galilei, capillary, Eötvös, and Morton numbers, have been proposed in this chapter indicating a simple relationship among density, viscosity, surface tension of fluids, and the scale ratio of physical model. This concept has been named SMDN (similitude of multiple dimensionless numbers). Subsequently, it was shown that some combinations of molten metals, such as Fe & Sn, Cu & Sn, Al & Ga, and Zn & Sn-Bi, could satisfy the strict relationship of physical properties required for the SMDN condition. As a matter of record, no one except for us has made the specific proposal of the multiple dimensionless numbers similitudes in the physical modelling of flow phenomena.

Physical modelling satisfying the SMDN condition can represent many kinds of flow phenomena influenced by inertial, gravitational, viscous, and surface tension forces as some examples of applications were shown in this chapter. This novel concept of physical modelling will enable precise, safe, and easy physical model experiments using low-temperature fluids to exactly represent flow phenomena in various high-temperature liquid operations within the pyro-metallurgical industries.

Nomenclature

V	Velocity of fluid,
g	Gravitational acceleration
L	Characteristic length
ρ	Density
η	Viscosity
σ	Surface tension
ν	Kinematic viscosity
i	σ/ρ
λ	Scale ratio of physical model
D	Diameter
α	Thermal diffusivity
β	Thermal expansion coefficient
ΔT	Temperature drop across convection layer

0	Subscript for prototype
b	Subscript for bubble
n	Subscript for nozzle

Author details

Yuichi Tsukaguchi[1]*, Kodai Fujita[1], Hideki Murakami[2] and Roderick I.L. Guthrie[3]

1 Steelmaking Research Lab., R&D Laboratories, Nippon Steel Corporation, Kamisu-City, Japan

2 R&D Laboratories, Nippon Steel Corporation, Futtsu-City, Japan

3 McGill Metal Processing Centre, McGill University, Montreal, QC, Canada

*Address all correspondence to: tsukaguchi.6fk.yuichi@jp.nipponsteel.com

References

[1] Pfister M, Chanson H. Two-phase air-water flows: Scale effects in physical modeling. Journal of Hydrodynamics. 2014;**26**:291-298. DOI: 10.1016/S1001-6058(14)60032-9

[2] Cho SM, Thomas BG. Electromagnetic forces in continuous casting of steel slabs. Metals. 2019;**9**: 471-438. DOI: 10.3390/met9040471

[3] Liu S, Yang X, Du L, Li L, Liu C. Hydrodynamic and mathematical simulations of flow field and temperature profile in an asymmetrical t-type single-strand continuous casting tundish. ISIJ International. 2008;**48**: 1712-1721. DOI: 10.2355/isijinternational.48.1712

[4] Saternus M, Merder T. Physical modelling of aluminum refining process conducted in batch reactor with rotary impeller. Metals. 2018;**8**:726-714. DOI: 10.3390/met8090726

[5] R-Lopez PE, Jalali PN, Bjorkvall J, Sjostrom J, Nilsson C. Recent developments of a numerical model for continuous casting of steel: Model theory, setup and comparison to physical modelling with liquid metal. ISIJ International. 2014;**54**:342-350. DOI: 10.2355/isijinternational.54.342

[6] Pieprzyca J, Merder T, Saternus M. Physical modelling of the process of mixing liquid metal in a tundish blown by gas. Meta. 2014;**53**:327-330

[7] Saternus M, Pieprzyca J, Merder T. Physical modelling of metallurgical processes. Materials Science Forum. 2016;**879**:1685-1690. DOI: 10.4028/www.scientific.net/MSF.879.1685

[8] Sahai Y, Emi T. Criteria for water modeling of melt flow and inclusion removal in continuous casting tundishes. ISIJ International. 1996;**36**:1166-1173. DOI: 10.2355/isijinternational.36.1166

[9] Mazumdar D, Guthrie RIL. The physical and mathematical modelling of continuous casting tundish systems. ISIJ International. 1999;**39**:524-547. DOI: 10.2355/isijinternational.39.524

[10] Moon K, Choi W, Isac M, Guthrie RIL, Nogami H. Physical and mathematical modeling in development of metal delivery system for single belt casting process. ISIJ International. 2003; **43**:1538-1547. DOI: 10.2355/isijinternational.43.1538

[11] T-Alonso E, Morales R, G-Hernandez S, P-Ramos J. Cyclic turbulent instabilities in a thin slab mold. Part I: Physical model. Metallurgical and Materials Transactions B: Process Metallurgy and Materials Processing Science. 2010;**41B**:583-597. DOI: 10.1007/s11663-010-9361-9

[12] Timmel K, Eckert S, Gerbeth G, Stefani F, Wondrak T. Experimental modeling of the continuous casting process of steel using low melting point metal alloys—the LIMMCAST program. ISIJ International. 2010;**50**:1134-1141. DOI: 10.2355/isijinternational.50.1134

[13] Timmel K, Wondrak T, Roder M, Stefani F, Eckert S, Gerbeth G. Use of cold liquid metal models for investigations of the fluid flow in the continuous casting process. Steel Research International. 2014;**85**: 1283-1290. DOI: 10.1002/srin.201300034

[14] Harada H, Toh T, Ishii T, Kaneko K, Takeuchi E. Effect of magnetic field conditions on the electromagnetic braking efficiency. ISIJ International.

2001;**41**:1236-1244. DOI: 10.2355/isijinternational.41.1236

[15] Tsukaguchi Y, Furuhashi S, Kawamoto M. Effect of magnetic field conditions on the electromagnetic braking efficiency. ISIJ International. 2004;**44**:350-355. DOI: 10.2355/isijinternational.41.1236

[16] Cramer A, Eckert S, Galindo V, Gerbeth G, Willers B, Witke W. Liquid metal model experiments on casting and solidification processes. Journal of Materials Science. 2004;**39**:7285-7294. DOI: 10.1023/B:JMSC.0000048743.43663.e9

[17] Li B, Lu H, Shen Z, Sun X, Zhong Y, Ren Z, et al. Physical modeling of asymmetrical flow in slab continuous casting mold due to submerged entry nozzle clogging with the effect of electromagnetic stirring. ISIJ International. 2019;**59**:2264-2271. DOI: 10.2355/isijinternational.ISIJINT-2018-866

[18] Martinez E, Maeda M, Heaslip LJ, Rodriguez G, Mclean A. Effects of fluid flow on the inclusion separation in continuous casting tundish. Transactions of the Iron and Steel Institute of Japan. 1986;**26**:724-731

[19] Singh S, Koria SC. Model study of the dynamics of flow of steel melt in the tundish. ISIJ International. 1993;**33**:1228-1237. DOI: 10.2355/isijinternational.33.1228

[20] Kurobe J, Iguchi M. Cold model experiment on melting phenomena of Zn ingot in hot dip plating bath. Materials Transactions. 2003;**44**:877-884

[21] Singh V, Dash SK, Sunitha S, Ajmani K, Das A. Experimental simulation and mathematical modeling of air bubble movement in slab caster mold. ISIJ International. 2006;**46**:210-218. DOI: 10.2355/isijinternational.46.210

[22] Zhang L, Yang S, Cai K, Li J, Wan X, Thomas BG. Investigation of fluid flow and steel cleanliness in the continuous casting strand. Metallurgical and Materials Transactions B: Process Metallurgy and Materials Processing Science. 2007;**38B**:63-83. DOI: 10.1007/s11663-006-9007-0

[23] Warzecha M. Numerical and physical modelling of steel flow in a one-strand continuous casting tundish. Meta. 2011;**50**:147-150

[24] Sun L, Li J, An L, Wang L, Tang H, Gao Y. Physical modeling on the mold flow field influenced by nozzle clogging. Cham. Pittsburgh: TMS: Proc. 8th Pacific Rim Int. Cong. on Advanced Materials and Processing; 2013. pp. 3163-3169

[25] Cloete JH, Akdogan G, Bradshaw SM, Chibwe DK. Physical and numerical modelling of a four-strand steelmaking tundish using flow analysis of different configurations. Journal of the Southern African Institute of Mining and Metallurgy. 2015;**115**:355-362

[26] Michalek K, Gryc K, Socha L, Tkadlečková M, Saternus M, Pieprzyca J, et al. The Quality of Online Health-Related Information – an Emergent Consumer Health Issue. Archives of Metallurgy and Materials. 2016;**61**:257-260. DOI: 10.1515/amm-2016-0048

[27] S-Campoy MM, Morales RD, N-Bastida A, C-Hernandez V, D-Pureco JC. A physical model to study the effects of nozzle design on dense two phase flows in a slab mold casting ultra-low carbon steels. Metallurgical and Materials Transactions B: Process Metallurgy and Materials Processing Science. 2017;**48B**:1376-1389. DOI: 10.1007/s11663-017-0918-8

[28] Mabentsela A, Akdogan G, Bradshaw S. Numerical and physical modelling of tundish slag entrainment in the steelmaking process. Journal of the Southern African Institute of Mining and Metallurgy. 2017;**117**:469-483

[29] Cwudziński A. Physical and mathematical modeling of bubbles plume behaviour in one strand tundish. Metallurgical Research & Technology. 2018;**115**:101-108. DOI: 10.1051/metal/2017081

[30] Zhang K, Liu J, Cui H, Xiao C. Analysis of meniscus fluctuation in a continuous casting slab mold. Metallurgical and Materials Transactions B: Process Metallurgy and Materials Processing Science. 2018;**49B**:1174-1184. DOI: 10.1007/s11663-018-1236-5

[31] Liu Y, Ersson M, Liu H, Jonsson PG, Gan Y. A review of physical and numerical approaches for the study of gas stirring in ladle metallurgy. Metallurgical and Materials Transactions B: Process Metallurgy and Materials Processing Science. 2019;**50B**:555-577. DOI: 10.1007/s11663-018-1446-x

[32] Jiang X, Cui Z, Chen M, Zhao B. Study of plume eye in the copper bottom blown smelting furnace. Metallurgical and Materials Transactions B: Process Metallurgy and Materials Processing Science. 2019;**50B**:782-789. DOI: 10.1007/s11663-019-01516-0

[33] Pieprzyca J, Merder T, Saternus M, Gryc K, Socha L. The influence of parameters of argon purging process through ladle on the phenomena occuring in the area of phase distributions: Liquid steel-slag. Archives of Metallurgy and Materials. 2019;**64**:653-658. DOI: 10.24425/amm.2019.127594

[34] Souza GM, Mendonça AFG, Tavares RP. Physical and mathematical modeling of inclusion behavior in a tundish with gas curtain. REM International Engineering Journal. 2020; **73**:531-538. DOI: 10.1590/0370-44672020730010

[35] Cwudziński A, Jowsa J, Gajda B, Hutny A. Physical modelling of fluids' interaction during liquid steel alloying by pulse-step method in the continuous casting slab tundish. Ironmaking & Steelmaking. 2020;**47**:1188-1198. DOI: 10.1080/03019233.2019.1708670

[36] Bielnicki M, Jowsa J. Physical and numerical modeling of liquid slag entrainment in mould during slabs casting. Metallurgical Research & Technology. 2020;**117**:509-12-11. DOI: 10.1051/metal/2020055

[37] Tsukaguchi Y, Fujita K, Murakami H, Guthrie RIL. Physical modelling for the precise representation of flow phenomena based on simultaneous similitude of multiple dimensionless numbers. Metallurgical and Materials Transactions B: Process Metallurgy and Materials Processing Science. 2021;**52B**:2726-2736. DOI: 10.1007/s11663-021-02226-2

[38] Tsukaguchi Y, Fujita K, Murakami H, Guthrie RIL. Physical modelling of flow phenomena based on simultaneous similitude of multiple dimensionless numbers. ISIJ International. 2021;**61**: 2897-2903. DOI: 10.2355/isijinternational.ISIJINT-2021-183

[39] Chanson H. The Hydraulics of Open Channel Flow. 2nd ed. Oxford: Butterworth-Heinemann; 2004 p. 257

[40] Clift R, Grace J, Weber M. Bubbles, Drops and Particles. New York: Dover Publications Incorporated; 2005. pp. 26-27

[41] Pfister M, Hager WH. History and significance of the morton number in

hydraulic engineering. ASCE Journal of Hydraulic Engineering. 2014;**140**: 02514001-025141-6. DOI: 10.1061/ (ASCE)HY.1943-7900.0000870

[42] Tesfaye F, Taskinen P. Densities of Molten and Solid Alloys of (Fe, Cu, Ni, Co) - S at Elevated Temperatures - Literature Review and Analysis. Helsinki: Aalto University Publications in Materials Science and Engineering; 2010. pp. 9-10. DOI: 10.13140/ 2.1.2804.1282

[43] Tanaka T, Hara S. Surface tension and viscosity of molten ferrous alloys. Materia. 1997;**36**:47-54 (in Japanese)

[44] National Astronomical Observatory of Japan. Chronological Scientific Tables. Tokyo: Maruzen; 2004. pp. 371-374 (in Japanese)

[45] Kasama A, Iida T, Morita Z. Temperature dependency of surface tension of molten metals. Journal of Japan Institute of Metal and Materials. 1976;**40**:1030-1038. (in Japanese). DOI: 10.2320/jinstmet1952.40.10_1030

[46] Gancarz T, Moser Z, Gasior W, Pstrus J, Henein H. A comparison of surface tension, viscosity, and density of sn and Sn–Ag alloys using different measurement techniques. International Journal of Thermophysics. 2011;**32**: 1210 1233. DOI: 10.1007/s10765 011 1011-1

[47] Assael MJ, Kalyva AE, Antoniadis KD. Reference data for the density and viscosity of liquid copper and liquid tin. Journal of Physical and Chemical Reference Data. 2010;**39**: 033105-0331-8. DOI: 10.1063/1.3467496

[48] Matsumoto T, Fujii H, Ueda T, Kamai M and Nogi K. Surface tension of molten metal using a falling droplet in a short drop tube. Transactions of JWRI 2005;34:29-33. Available from: http:// www.jwri.osaka-u.ac.jp/publication/ trans-jwri/pdf/342-29.pdf [Accessed: 12 March 2017]

[49] Assael MJ, Armyra IJ, Brillo J, Stankus SV, Wu J, Wakeham EA. Reference data for the density and viscosity of liquid cadmium, cobalt, gallium, indium, mercury, silicon, thallium, and zinc. Journal of Physical and Chemical Reference Data. 2012;**41**: 033101-031-16. DOI: 10.1063/1.4729873

[50] Hogness TR. The surface tensions and densities of liquid mercury, cadmium, zinc, lead, tin and bismuth. Journal of the American Chemical Society. 1921;**43**:1621-1628. DOI: 10.1021/ja01440a026

[51] Dinsdale AT, Quested PN. The viscosity of aluminium and its alloys—A review of data and models. Journal of Materials Science. 2004;**39**:7221-7228

[52] Leitner M, Leitner T, Schmon A, Aziz K, Pottlacher G. Thermophysical properties of liquid aluminum. Metallurgical and Materials Transactions A: Physical Metallurgy and Materials Science. 2017;**48**:3036-3045. DOI: 10.1007/s11661-017-4053-6

[53] Bainbridge IF, Taylor JA. The surface tension of pure aluminum and aluminum alloys. Metallurgical and Materials Transactions A: Physical Metallurgy and Materials Science. 2013;**44**:3901-3909. DOI: 10.1007/s11661-013-1696-9

[54] Plevachuk Y, Sklyarchuk V, Gerbeth G, Eckert S. Thermophysical properties of liquid tin–bismuth alloys. International Journal of Materials Research. 2010;**101**:839-844. DOI: 10.3139/146.110357

[55] Luty P, Pronczuk M. Determination of a bubble drag coefficient during the

formation of single gas bubble in upward coflowing liquid. PRO. 2020;**8**:0999-1-19. DOI: 10.3390/pr8080999

[56] Tomiyama A, Kataoka I, Zun I, Sakaguchi T. Drag coefficients of single bubbles under normal and micro gravity conditions. JSME International Journal Series B Fluids Therm Engineering. 1998; **41**:472-479. DOI: 10.1299/jsmeb.41.472

[57] Mori Y, Hijikata K, Kuriyama I. Experimental study of bubble motion in mercury with and without a magnetic field. Journal of Heat Transfer. 1977;**99**: 404-410. DOI: 10.1115/1.3450710

[58] Saylor JR, Bounds GD. Experimental study of the role of the weber and capillary numbers on Mesler entrainment. AICHE Journal. 2012;**58**: 3841-3851. DOI: 10.1002/aic.13764

[59] Iwase Y, Honda S. An interpretation of the Nusselt-Rayleigh number relationship for convection in a spherical shell. Geophysical Journal International. 1997;**130**:801-804. DOI: 10.1111/j.1365-246X.1997.tb01874.x

Section 3

Gravity Die Casting

Chapter 5

Casting Techniques: An Alternative for Producing Parts with Recycled Al in the Gravity Die Casting Process

Narducci Carlos Jr.

Abstract

This work applied the grain refinement technique by heterogeneous nucleation and precipitation hardening to investigate the effect of size and morphology of β-Fe particles on Al-Si alloys' mechanical behavior Fe-critical, inoculated via Nb+B and heat-treated. The samples for the microstructural analyses were produced according to the standard mold, Test Procedure-1 (TP-1) and, analyzed by optical microscope with polarised light and filter plate and differential interference contrast (DIC) and by X-ray energy dispersive spectroscopy (XRD SEM) with EDS detector analyzer. The specimens for the mechanical tests were cast in a metal mold according to ASTM B108. The combined effect of manipulating the studied alloy Al10Si1Fe0.35Mg resulted in reduced and spheroidized β-Fe precipitates with improved mechanical properties in the material. Properties are similar to those achieved by commercially used alloys with engineering applications in structural and safety parts.

Keywords: recycled Al, Al-Si, Nb + B, casting, heat treatment, mechanical properties, grain refinement, and intermetallic precipitates

1. Introduction

The increasing need for weight reduction in means of transport has made Aluminum (Al) one of the main consumer materials in the world. Another benefit of Al is its capacity to be recycled infinite times without losing its properties. Aluminum recycling is a way of reusing the material used, transforming it into new products for consumption. Its contribution in reducing the energy spent in its production and favor of the environment has become an urgent issue, of all, with a consensus that the use of the planet's resources must be sustainably [1, 2].

However, recycled Al is still restricted for structural and safety products due to the percentage of Fe-critical contained in Al alloys coming from recycling.

The Fe-critical precipitates the formation of coarse particles of β-Al5FeSi (β-Fe), which weaken the material, impairing its mechanical properties [3]. Therefore, in the search for a new perspective for the use of recycled Al in gravity die casting processes for the production of structural and safety parts and components, this work studies the creation of a strengthening mechanism in the structure of the material through the manipulation of the intermetallic precipitates of β-Fe by the heterogeneous nucleation phenomenon, through the addition of the inoculant via niobium (Nb) and boron (B) and, the addition of the element magnesium (Mg), to cause precipitation hardening. The simultaneous use of the two techniques results in improvements to the material's mechanical properties.

Figure 1 shows some examples of products with automotive applications that can benefit from this process: Master Cylinders, Tweezers, camshaft rocker arms, and suspension bracket components, among other automotive and aerospace components.

1.1 Nucleation

From classical nucleation theory, we know that nucleation of a particle can occur in a homogeneous or heterogeneous manner, with the formation of nuclei that can form an embryo and eventually be able to grow, depending on the temperature or variation of the free energy of the system [4–7]. An understanding of homogeneous nucleation will help us to understand heterogeneous nucleation better. In homogeneous nucleation, there is no interface in the system and occurs when there is a grouping of solute atoms in the matrix so that the radius of the particle embryo exceeds a certain radius value, known as the critical radius (r^*), for the formation of a stable nucleus, given a certain temperature fluctuation. Solid nuclei emerge from the liquid. The embryo becomes a stable nucleus from the nucleation process which is energetically activated when the particle exceeds the critical radius, representing the free energy peak (ΔG). **Figure 2** brings a schematic representation of the homogeneous nucleation process.

Figure 1.
Customers and automotive products with the potential use of the techniques.

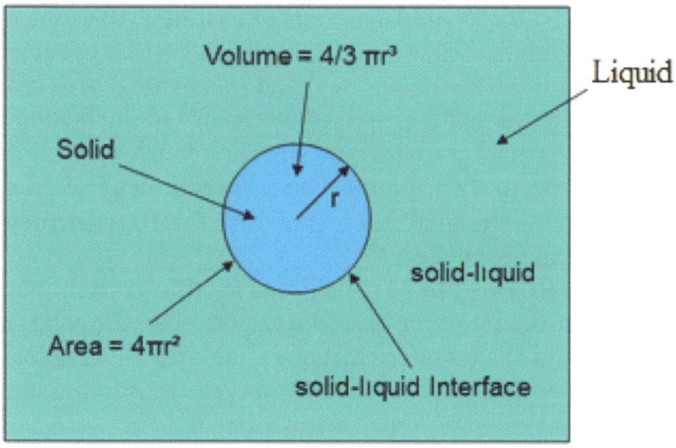

Figure 2.
Schematic illustration of the homogeneous nucleation particle, adapted from [7].

The theory is based on the free energy between the liquid and solid interfaces. The highest free energy for homogeneous nucleation to occur (ΔG* hom) can be expressed as shown in Eq. (1) [4]:

$$\Delta G *_{\text{hom}} = -V\,\Delta G_v + A\gamma_{SL} \qquad (1)$$

Where: V is the volume of the sphere, ΔG is the variation of the free energy of the system, A is the area of the sphere, and γSL is the surface tension between the solid and the liquid [4].

For heterogeneous nucleation, there must be an interface (particles of other elements, impurities, etc.), a phenomenon that decreases the variation of free energy for forming an embryo. Therefore the nucleation will be considered heterogeneous when it takes place on the surface of the mold or of the particles present in the system. The theory was developed by [8]. The critical factor is the wetting angle (θ), which occurs as a function of the interface energy of the mold surface γSL, the crystal-substrate interface γSM, and the liquid-substrate interface γML, as shown in **Figure 3**. The wetting angle θ can be calculated as shown in Eq. (2) [8]:

$$\cos\theta = \left(\gamma_{ML} - \gamma_{SM}\right) / \gamma_{SL} \qquad (2)$$

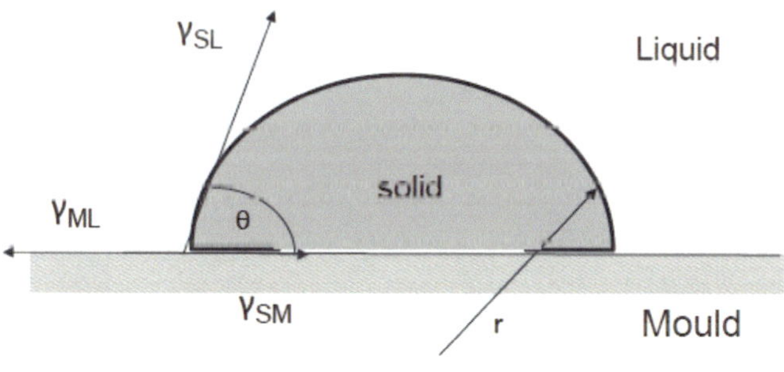

Figure 3.
Schematic illustration of the wetting angle between the mold and the solid, adapted from [7].

One of the most important technical aspects of casting is the control of nucleation grain size and shape, together with the segregation effects that occur during casting, which can significantly influence the physical and mechanical properties of the cast product. Most commonly, control is exercised by the use of nucleating agents (inoculants). With this, the growth of grains occurs simultaneously with the nucleation of other grains in different parts of the liquid. The competition between nucleation and grain growth will determine the final macrostructure (grain refinement), which gives the product its mechanical properties.

After nucleation, the first step to be considered is grain growth. The grain growth depends on the nature of the interface between the wetting, the solid, and the liquid. The structure and shape of the interface influence both the morphology and size of the grains and the number and distribution of imperfections in the solid. The solidification can occur in planar, cellular, or dendritic form [9].

In Al alloys, grain growth is dendritic [10]. Suppose a state of metastability or instability is created near the growth interface by the occurrence of an inverted gradient of free energy. In that case, the growth energy will break and grow laterally and develop lateral ramifications, according to **Figure 4**.

Scientists developed the grain refinement theory from observations with Al-Ti-B additions in Al alloys. Currently, two strands are used to explain these phenomena; (a) the nucleating paradigm with the nucleating particle theory and phase diagram theory and (b) the solute paradigm theory [12, 13]. The schematic presentation in **Figure 5** brings the proposal presented for the theories.

With the knowledge of grain refinement, the studies will delve into manipulating the available parameters to obtain the best result of grain refinement. With this, the element Nb emerged [14–17].

1.2 Precipitation hardening

Precipitation hardening is a technique used in Al-Si cast alloys, usually with a Fe content within recommended limits (e.g., ASTM A357 alloy), to strengthen the mechanical properties of the material. The process consists of adding a precipitating

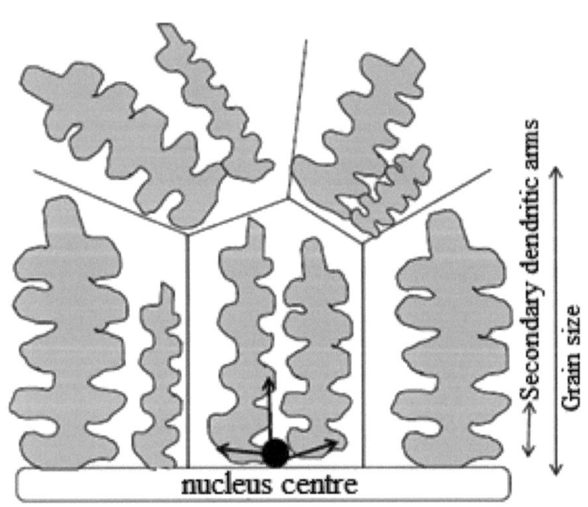

Figure 4.
Schematic illustration of dendritic grain growth, adapted from [11].

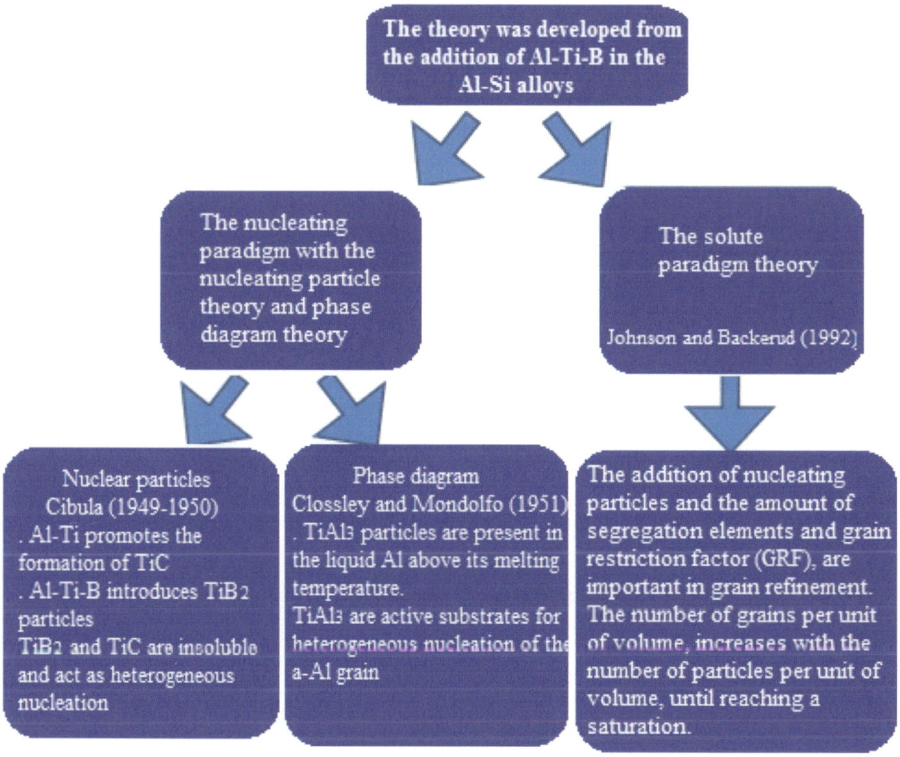

Figure 5.
Schematic presentation of the theories: Nucleating particles, phase diagram and solute paradigm [12, 13].

alloying element, such as Mg or Cu, to the material, with subsequent solubilization and precipitation heat treatment (T6). During the heat treatment, the material is heated within the single-phase region for a sufficient time to solubilize the solute atoms, followed by rapid cooling to obtain a supersaturated solid solution. Subsequently, the alloy is reheated to a temperature below the single-phase region to allow the precipitation of finely dispersed particles from the supersaturated solid solution. The formation of a dispersion of fine precipitates hinders the movement of the disagreements [7, 18–21].

Heat treatment is one of the mechanisms that, through phase transformation, can change the microstructure of the material, giving it a set of desirable mechanical characteristics. The heat treatment route is carried out, having as basis the phase diagram, in a convenient way to achieve some of these desired transformations, related to time and temperature, to find the ideal processing parameters for a given alloy. Most phase transformations depend on the reaction progress as a function of time or the transformation rate involving solid phases. A new phase is formed during transformations, with different physical and chemical characteristics from those of the phase that gave rise to it. Normally, the formation of new phases begins by forming small nuclei with subsequent growth. The growth will continue until the equilibrium condition is reached. This heat treatment, applied to Al alloys, is known as precipitation hardening because the small particles of the new phase are known as "precipitates." The denomination "age hardening" is also used to designate this procedure because the resistance develops over time. **Figure 6** helps to understand the mechanism of the transformations [7, 22].

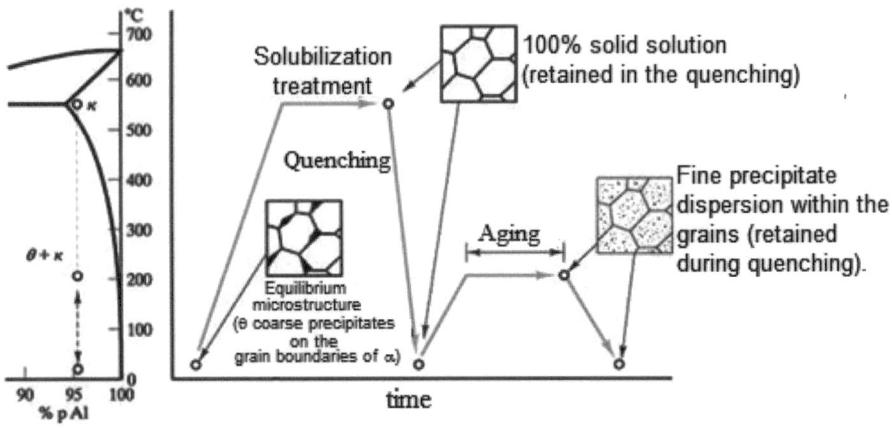

Figure 6.
Heating up to solubilization temperature, rapid cooling (quenching), and then reheating below solubilization temperature (aging). With this, a fine dispersion of precipitates is formed within the grains, adapted from [22].

Al-Si alloys, with the addition of some more elements, such as Mg or Cu, are heat-treatable alloys, i.e., they can have their mechanical strength increased by appropriate heat treatment. The mechanical properties of Al-Si alloys depend on the morphology and distribution of the eutectic second phase particles. The brittleness of the coarse, second phase fibers is the main reason for low elasticity and low tensile and impact strengths. However, the fine, dispersed, and globular second phase particles, combined with precipitates formed during the aging process, can result in excellent mechanical properties. Significant improvements in the mechanical properties of alloys, especially elasticity, can be obtained with solubilization and aging treatment, which changes the morphology and distribution of the precipitated particles. Previous studies show that alloys aged with high silicon content, above 5%, have better mechanical properties than alloys with low silicon content [23]. The choice of the element used for precipitation in Al-Si alloy in this work was Mg.

2. Experimental procedures

2.1 Casting

The materials used were: Primary Al (supplied by HYDRO), Si (supplied by LIASA), Fe (supplied by MEXTRAMETAL), and Al4Nb0.05B master alloy (supplied by Companhia Brasileira de Metalurgia e Mineração - CBMM). The casting materials (supplied by ALFA TREND).

Four types of specimens (CDPs) were produced, from the Al10Si1Fe alloy: without and with the addition of the inoculant NbB, to allow the evaluation of the influence of grain refinement on the morphology of the β-Fe precipitates; and, without and with the addition of the element Mg, to enable the evaluation of the aging potential by precipitation of the alloy with the introduction of this element. **Table 1** presents the chemical composition of each batch produced.

The CDPs were produced at the Foundry (Sunny in SP-Itaquaquecetuba). The metal melting was in an electric crucible furnace, with a capacity of 60 kg of material.

Batch	Elements (Wt. %)					
	Al	Si	Fe	Nb	B	Mg
1°	balance	10	1	0	0	0
2°	balance	10	1	0,05	0,00625	0
3°	balance	10	1	0,05	0,00625	0,35
4°	balance	10	1	0	0	0,35

Table 1.
Chemical composition of the CDPs.

After adding each element, the temperature was stabilized at 850 ± 10°C, a retention of 1 hour was applied to ensure complete dissolution. Degassing was then performed by the addition of hexachloroethane tablets. At each sample, collection homogenization was achieved by 30 s of manual stirring and a further stabilization in temperature at 720 ± 10°C, followed by a second homogenization at the same temperature, with 30 s of stirring and sampling. The verification of the base alloy was by atomic absorption spectrometry.

Figure 7 shows the casting, through the gravity die casting process, using a 1020 steel metal mold manufactured according to ASTM B108 (supplied by Alpha Trend). The mold construction followed the international standard, enabling comparisons with other studies and research results. The mold was built in two halves and two cavities, producing two CDPs per pour. The mold feeding system was made with a central filling channel that distributes the metal in two auxiliary side feeding channels, making the liquid metal feed the product from bottom to top and through two inlet channels in each part. This system minimizes the material turbulence during the part filling. Before pouring the liquid metal, the mold was painted with water-based graphite paint and heated with a blowtorch to a temperature of 250°C.

2.2 Microscopic analysis

Figure 8 shows the location from where the samples were taken from the body of the CDPs and cut in the horizontal and vertical directions when filling the piece.

For grain size analysis, the surfaces of the cross-section of the samples were prepared with 2400 mesh sandpaper, without polishing, according to **Figure 9**, in sequence the samples were attacked for 15 seconds with Poulton's acid solution (60% HCl at 37%; 30% HNO_3 at 65%; 5% HF at 50% and 5% H_2O), then washed in water for 20 seconds. They were bleached with (67% HNO_3 at 65%, 5% HF at 50%, and 13% H_2O) and a time of 15 seconds. For the analysis of microstructural constituents, the cross-section surfaces, after prepared with 2400 mesh sandpaper, were polished with 1 μm alumina suspension and then chemically attacked with Keller reagent (95.0 ml distilled water, 2.5 ml HNO_3, 1.5 ml HCl and 1.0 ml HF).

The macro and microstructures were examined in an optical microscope with polarized light, filter plate, and differential interference contrast (DIC). The measurement of the average grain size (G) was conducted by the linear intercept method according to the ASTM E112-10 standard, 1996. The β-Fe intermetallic precipitate was analyzed by X-ray Dispersive Energy Spectroscopy (EDS) with the aid of a Tescan scanning electron microscope from Oxford Instruments to obtain the spatial spectra Al, Si, and Fe.

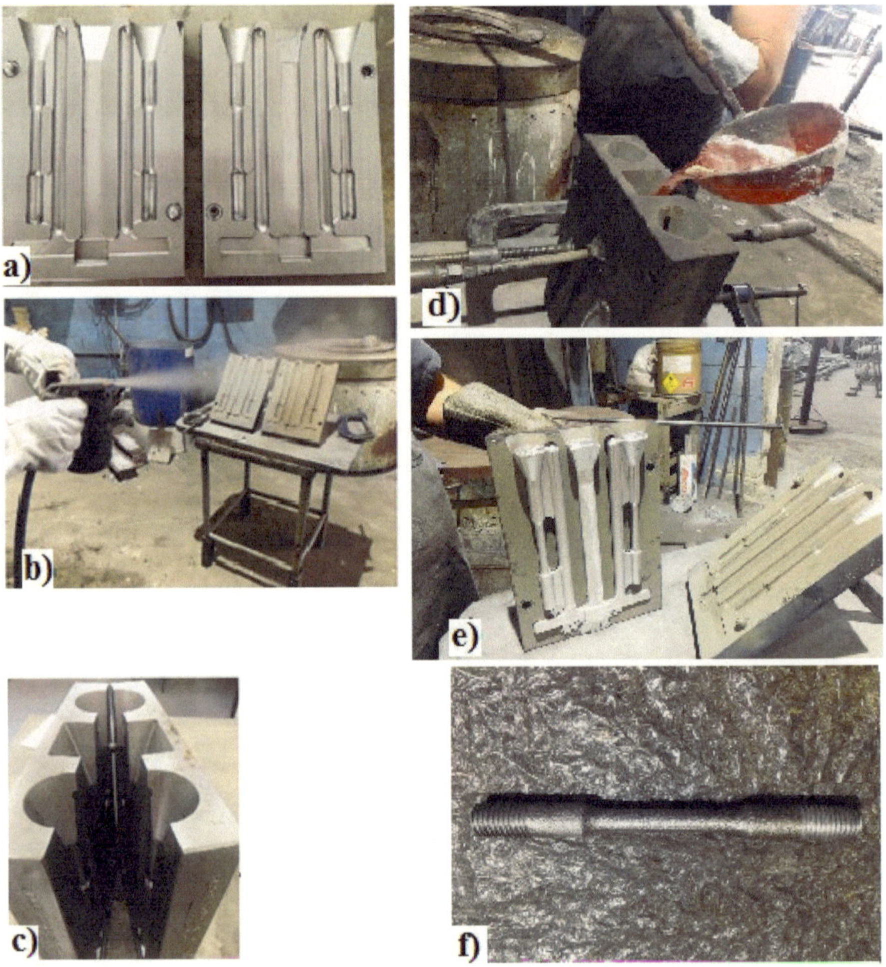

Figure 7.
Leaking of the CDPs. a) Open mold. b) Painting the mold. c) Closing the mold. d) Leaking with manual transportation by ladle. e) Open mold with the CPD's. f) CPD without channels and massalots.

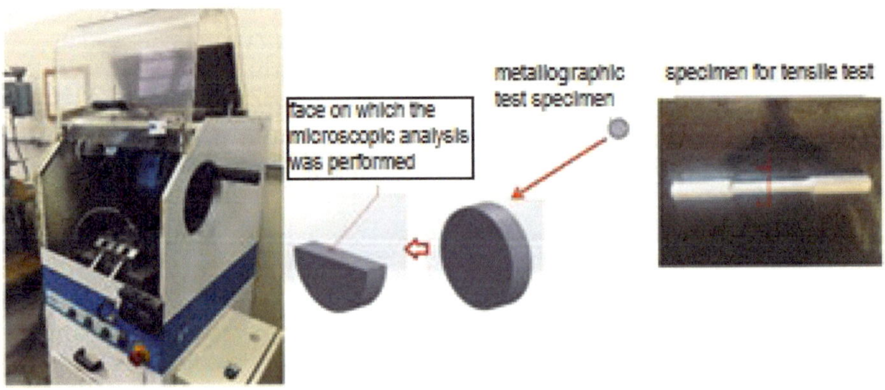

Figure 8.
Cutting of the CPDs for sample preparation.

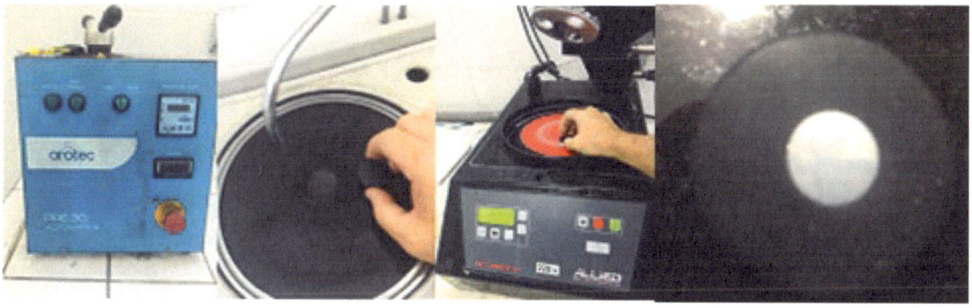

Figure 9.
Mounting, sanding, polishing, and finished sample.

2.3 Heat treatment

The heat treatment (T6) was performed in a Mufla furnace, OTTO WOLPERT - WERKE, located in the Metallurgy Laboratory of ITA. The parameters of heat treatment used were: temperature of 525 ± 5°C, time of 4 h ± 15′ (solubilization), immediately transferred to a water tank at room temperature (25°C) for 15 minutes (quenching). Subsequently, the CDPs were heated at a temperature of 175 ± 5°C, time of 7 h ± 15′ (precipitation or artificial aging), with subsequent cooling in the open air, as shown in **Figure 10**.

2.4 Tensile test

The tensile test was performed in a universal machine (brand Quanteq), model Emic Trd28 and equipped with a TestScript304 software for testing methods, located in the metallurgy laboratory of the Federal Institute of São Paulo (IFSP) - Campus Itaquaquecetuba, as shown in **Figure 11**. Data on yield stress, maximum stress, and material elasticity were obtained in the tests. Five CDPs were tested for each batch of material (**Table 1**), with and without T6.

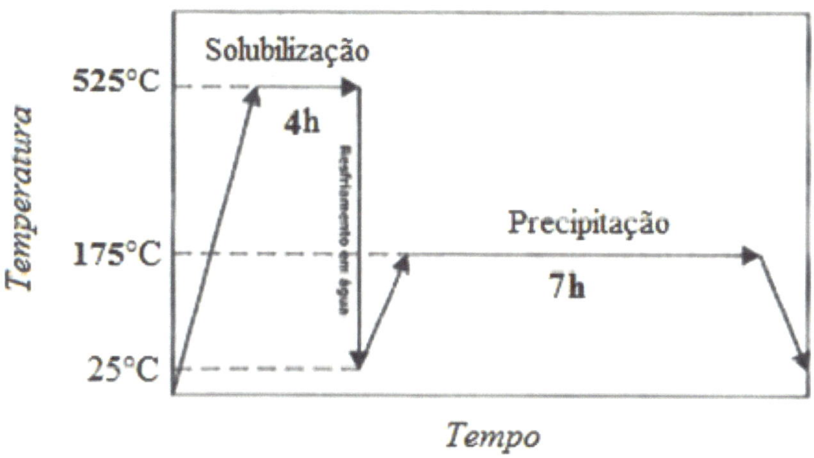

Figure 10.
Solubilization treatment followed by rapid cooling and precipitation.

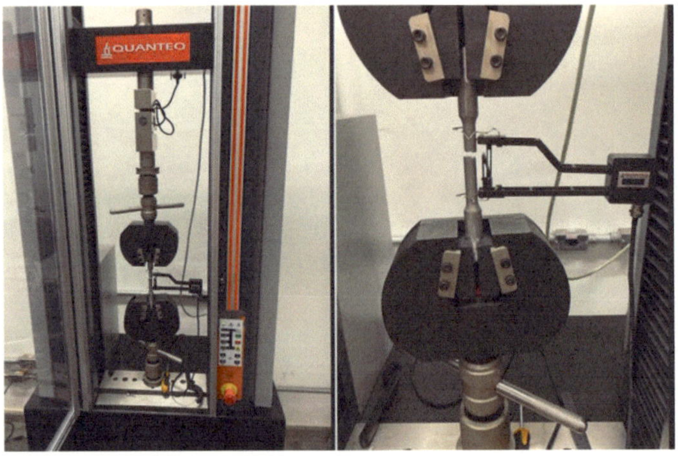

Figure 11.
Tensile test of sample Al10Si1Fe with the addition of the inoculant Nb + B.

3. Results and discussion

In **Figure 12**, the microstructural analyses of the material revealed a drastic reduction in the grain size of the samples without and with the addition of the Nb + B inoculant.

Highlighted in white outlines the dendrites without and with the addition of the inoculant Nb + B. The average grain size was calculated according to the linear intercept method. Overall, the reduction in the average grain size was 656% compared to the base alloys without the inoculant addition (from 787 μm to 120 μm).

This transformation was due to the underlying mechanism of heterogeneous nucleation of clusters of Nb aluminum niobite substrates (Al_3Nb) and Nb niobium borides (NbB_2). These substrates were found in the core of α-Al grains, as shown in **Figure 13**. The heterogeneous transformation mechanism is also demonstrated in the literature and presented as shown in Eqs. (1) and (2) [24].

$$Nb + 3Al => Al_3Nb \tag{3}$$

$$Al3Nb + AlB2 => NbB_2 + 4\ Al \tag{4}$$

With the microstructural analyses, one can also identify the drastic change in the morphology of the β-Fe particles. **Figure 14** shows by SEM images of the spectra of β-Fe highlighted by light gray color, being in the samples without addition of the inoculant of elongated shape, **Figure 14(a)** and **(b)** and with the addition of 0.05% of Nb + B with reduced size and spheroidised shape, **Figure 14(c)** and **(d)**.

The analysis of **Figure 14** shows the drastic reduction in the size and morphology of the β-Fe intermetallic precipitates. Practically, the extinction of microporosity points, which are extremely harmful to the material, are crack initiation points when the material is subjected to mechanical stress. It is worth mentioning that this drastic reduction in the size of the intermetallic precipitate structures was achieved with a slow cooling rate. With this condition, it is very difficult to obtain the 'modification' of the needle-shaped eutectic phase and thick plate-shaped intermetallic precipitates

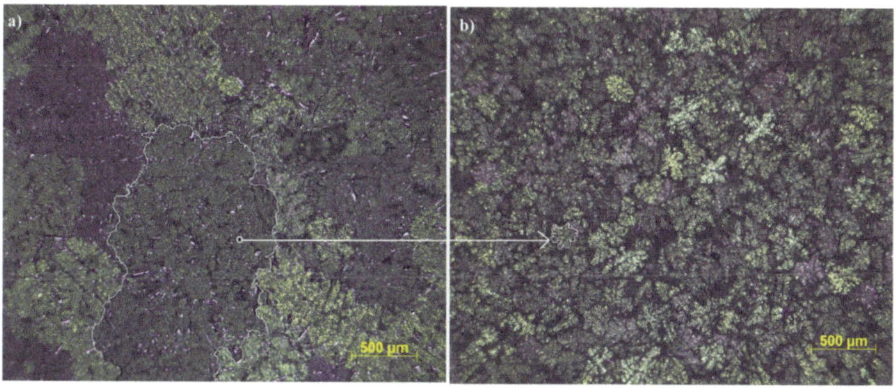

Figure 12.
Al10Si1Fe alloy, a) without inoculant and b) with Nb + B inoculant.

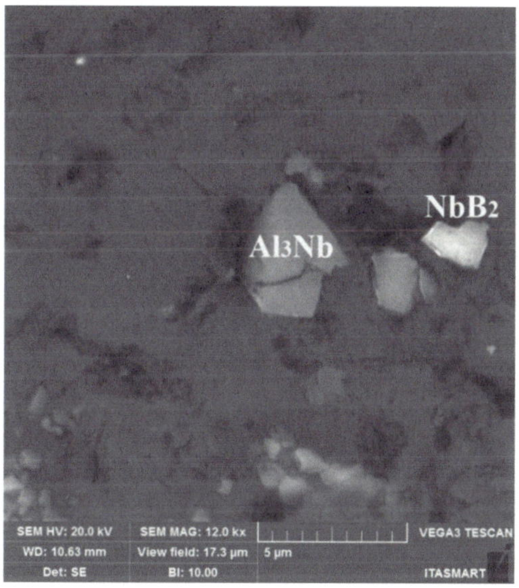

Figure 13.
Aluminum niobide substrates of (Al_3Nb) and niobium borides (NbB_2).

to morphology with reduced shape and size and spheroidized. This condition is highly desirable in the structure of the material.

After heat treatment (T6), the CDPs were tested in a universal traction machine under the conditions studied, as shown in **Table 1**. The results were tabulated, and the mean values of each sample were shown in the graph with the stress–strain curves, as shown in **Table 2** and **Figure 15**. Each color represents a test condition, as follows:

Red: samples with the base material.

Lilac: samples without the inoculant Nb + B, with Mg and heat treatment.

Green: samples with the addition of the inoculant Nb + B, without adding the element Mg and without heat treatment.

Blue: samples with the addition of the inoculant Nb + B, without adding the element Mg and with heat treatment.

Gray: samples with the inoculant Nb + B, with Mg and heat treatment.

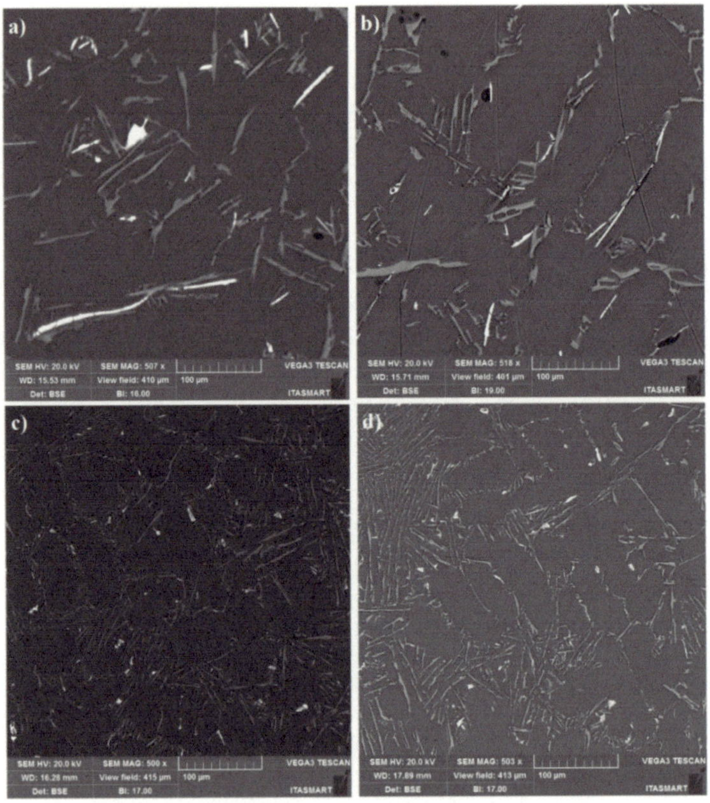

Figure 14.
Image with 500X magnification, showing the morphology of the β-Fe spectra in light gray color. (a, b) before addition of Nb + B; (c, d) after addition of Nb + B.

	Propriedades mecânicas		
	Tensão de escoamento (MPa)	**Tensão máxima (MPa)**	**Elasticidade (%)**
	95.5	197	2.12
	143	191	2.60
	194	261	2.02
	77.5	194	7.56
	208	300	4.66

Table 2.
Mechanical properties for each sample condition studied.

When we compared the red curve with the purple curve, the results were: gain of 50% in the yield strength (from 95.5 MPa to 143 MPa), without a gain in the strength limit (from 197 MPa to 191 MPa), with little evolution regarding the percentage elongation (from 2.12% to 2.60%). In the alloy without the addition of the inoculant Nb + B, the precipitates of β-Fe coming from the Fe-critical in the alloy present with morphology in the form of plates or coarse needles, as shown in **Figure 14(a)** and **(b)**, which are generally not affected by the typical T6 heat-treatment process. In the alloy

DOI: http://dx.doi.org/10.5772/intechopen.99983

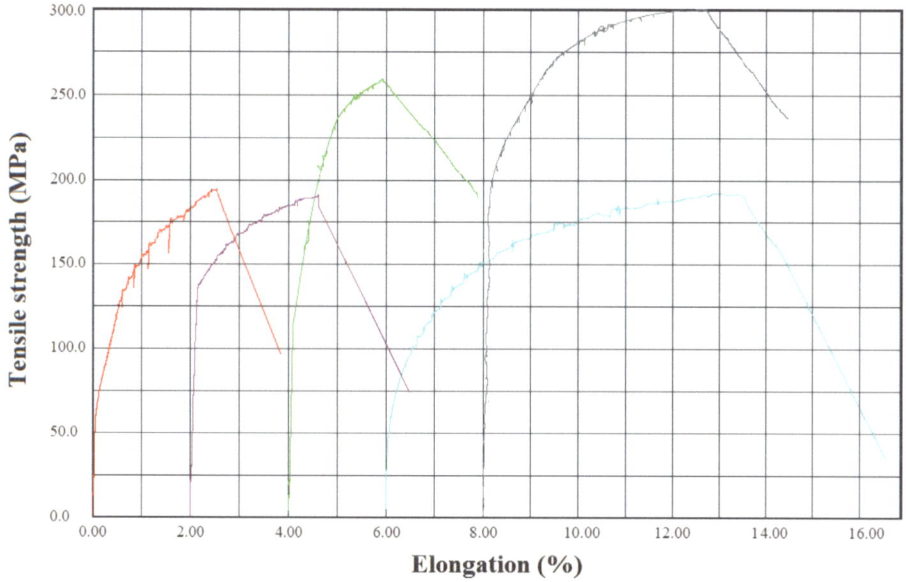

Figure 15.
A resistência à tração em função do alongamento Para os lotes fundidos. Material base (curva vermelha); com adição do elemento Mg e tratamento térmico (curva lilás); com adição de Nb + B e sem tratamento térmico (curva verde); com adição de Nb + B e com tratamento térmico (curva Azul); com adição de Nb + B, o elemento Mg e T6 (curva cinza).

with Mg, Mg_2Al_3 precipitates are formed, solubilized, and precipitated in finely dispersed particles during the T6 heat treatment, serving as a reinforcement mechanism for the material in its elastic region, which justifies the increase in elasticity found. On the other hand, in the plastic region, with the increase in stress, the β-Fe particles are like rigid points in the material's structure. They did not deform, initiating cracks, which ended up embrittling and limiting the maximum strength of the material. Signaling that the effect of heat treatment is not beneficial in alloys with Fe-critical.

When we compare the red curve with the green curve, the results were: gain of 203% in the yield strength (from 95.5 MPa to 194 MPa), a gain of 32.5 in the resistance limit (from 197 MPa to 261 MPa). However, without evolution as for the percentual elongation (from 2.12% to 2.02%). The reduced grains due to the inoculation via Nb + B brought greater grain contours. The grain contours are discontinuities that hinder the movement of discordances, improving the strength of the material. The α-Al grain inoculation via Nb + B caused a casting structure with equiaxial and reduced grain size and a structure without coarse plates of β-Fe particles and no presence of micro porosities, as verified in **Figure 14(c)** and **(d)**. However, the smaller and stiffer grains ended up reducing the elongation of the material, being below the reference value of the base alloy. Signaling that only the addition of the inoculant Nb + B was not enough to achieve the desired elongation to the material with Fe-critical.

When we compared the red curve with the blue curve, the mechanical behavior regarding the percentage elongation was transformed. The gain was 357% (from 2.12% to 7.56%). However, there was a reduction in the yield strength (from 95.50 MPa to 77.5 MPa) and no evolution in the strength limit of the material (from 197 MPa to 194 MPa). Due to the heat treatment, solubilization, the α-Al grain size was reconstituted. The plastic deformation started at a lower stress plateau but extended until about 7.6% strain. The material's behavior regarding the strength limit remained similar to that of the base material. Similar to the previous discussion, the material without the coarse plates formed by the β-Fe particles and without the presence of micro porosities improved its ductility, presenting a higher plastic deformation than the other conditions, with the retardation of the crack points, promoting a continued strain hardening until reaching the maximum stress during the plastic deformation. Although these characteristics are interesting for some mechanical applications, the addition of Nb + B, with subsequent T6 heat treatment, did not affect raising the desired strength levels without the addition of the Mg element.

The Mg element was introduced, and the aging treatment was carried out to achieve higher strength levels.

With the addition of the element Mg in the alloy refined with Nb + B and after the heat treatment of solubilization and precipitation T6 (gray curve), there were gains in tensile strength and elasticity levels when compared to the base alloy. Regarding the elongation, there was a gain of 90% compared to the alloy base material (from 2.12% to 4.66%), in the yield strength, the gain was 217.5% (from 95.5 MPa to 208 MPa), and in the tensile strength, the gain was 53% (from 197 MPa to 300 MPa).

It is interesting to note that only introducing the Mg element and the aging precipitation treatment (T6) increase the yield strength. Still, it is not efficient to raise the strength limit or ductility of the alloy. The introduction of only Nb + B as inoculant significantly increases the yield strength, raises the strength limit, but there is a 5% loss in elongation. When applying to this alloy, which received only the inoculant Nb + B, the aging treatment significantly increases the elongation (357%) but with loss in yield strength. The best combination of properties was observed when, in addition to the inoculant Nb + B, the element Mg was incorporated into the alloy. The aging treatment was performed (T6). In this case, there was an increase in yield strength (217.5%) in the strength limit (53%) and elasticity, measured by elongation (90%).

4. Conclusions

The addition of the inoculant Nb + B is a powerful tool for the primary α-Al grain refinement in AlSi alloys with critical Fe.

The refinement of the α-Al grain from the proposed inoculant (Nb + B) significantly changed the morphology and size of β-Fe precipitates, making them reduced and spheroidized.

In the Al10Si1Fe cast alloy, inoculated with 0.05% Nb and 0.063% B, with aging (T6) and without introducing the Mg element, there is a large gain in elongation (357%), but without gains in mechanical strength.

In the Al10Si1Fe cast alloy, inoculated with 0.05% Nb and 0.063% B, with aging (T6) and with the introduction of the Mg element, it was the best condition of the mechanical property achieved, with an increase in yield strength (217.5%), strength (53%) and elasticity, measured by elongation (90%).

The refinement promoted by the addition of Nb + B inoculant did not change the precipitation hardening mechanism of Mg2Al3. Similarly, the reconstitution of the α-Al grain size promoted during the solubilization of the material did not alter the morphology of the reduced and spheroidised β-Fe particles.

The beneficial actions of the two techniques used for strengthening the material's mechanical properties added up, providing strength and elasticity to the material, properties that are sought after in engineering applications.

The mechanical properties obtained with the studied alloy suggest using alloys coming from recycling (Fe-Critical) in engineering applications with structural and safety parts.

Acknowledgements

The author would like to thank CAPES (Coordenação de Aperfeiçoamento de Pessoal de Nível Superior) - funding code 001, for providing financial support for this study. The author is also grateful to the Institute of Aeronautical Technology, the Institute of Advanced Studies, and the Federal Institute of Education, Science and Technology of São Paulo (IFSP) - Itaquaquecetuba campus for the practice and support very kindly provided.

Conflict of interest

The authors declare no conflict of interest.

Notes/thanks/other declarations

Thanks to the teacher Dr. A. J. Abdalla, for every support during this research.

Author details

Narducci Carlos Jr.[1,2,3]

1 Aeronautics Institute of Technology – ITA, São José dos Campos, Brazil

2 Institute for Advanced Studies – IEAv, São José dos Campos, Brazil

3 Federal Institute of São Paulo – IFSP, Itaquaquecetuba, Brazil

*Address all correspondence to: cnarducci@ifsp.edu.br

References

[1] GREEN J.A.S. Aluminum Recycling and Processing for Energy Conservation and Sustainability; ASM International Materials Park, Ohio 44073-0002; US, 2007.

[2] MACHADO, C. T. S. et. al. A reciclagem de alumínio como vantagem estratégica de negócios em uma indústria metalúrgica. estudo de caso; XI INC; XI EPG; Universidade do Vale do Paraíba - 2011.

[3] TAYLOR, J. A; Iron-containing intermetallic phases in Al-Si based casting Alloys. ELSEVIER, Procedia Materials Science 1 19 – 33, Australia, 2012. CALLISTER, W. D. J. Fundamentos da Ciência e Engenharia de Materiais. 2ª Ed. LTC. Brasil, R. J. 2005.

[4] PORTER, D. A. EASTERLING, K. E. Phase transformations metals and alloys. Chapman Halls. Van Nostrand Reinhold Co., UK, London, 1981.

[5] SHEWMON, P.G. Transformations in metals, McGrawHill. New York, 1969.

[6] REED-HILL, R. E. Princípios de Metalurgia Física. Second Edition; Editora Guanabara. R.J, 1982.

[7] CALLISTER, W. D. J. Fundamentos da Ciência e Engenharia de Materiais. 2ª Ed. LTC. Brasil, R. J. 2005.

[8] TURNBULL, D. FISHER, J. C. Journal of Chemical Physic. 17.71 (1948): 71-73. 1948.

[9] JACKSON K, A. HUNT, J. P. Acta Met. 13(1965): 1212. 1965.

[10] TRIVERDI, R. KURZ, W. Int. Mat. Rev. 39 (1994): 129. 1994.

[11] BRUNATTO, S. F. Ligas Metálicas e Diagramas de Fases. INTMAT-2, GTFAP&MP, CNPQ. São Paulo, 2012.

[12] MARK, E. DAVID, ST. J. Grain Refinement of Aluminum Alloy: Part I. The Nucleant and Solute paradigms – A Review of the Literature. Metallurgical and materials Transactions. Jun 1999; 30A, 6; Academic Research library, pg 1613. Australia, 1999.

[13] MARK, E. DAVID, ST. J. Grain Refinement of Aluminum Alloy: Part II. Confirmation of, and a mechanism for, the Solute paradigms. Metallurgical and materials Transactions. Jun 1999; 30A, 6; Academic Research library, pg 1625. Australia, 1999.

[14] NARDUCCI, C. J. BROLLO, G. L. SIQUEIRA, R. H. M. ANTUNES, A. S. ABDALLA, A. J. Effect of Nb addition on the size and morphology of the β-Fe precipitates in recycled Al-Si alloys; Scientific Reports, Springer Nature. https://doi.org/10.1038/s41598-021-89050-5. Brazil, SP, 2021.

[15] NOWAK, M. BOLZONI, L. BABU, N. H. Grain refinement of Al-Si alloys by Nb-B inoculation. Part I. Materials and Desigs. 66 (2015) 366-375. UK, 2015.

[16] BOLZONI, L. NOWAK, M. HARI BABU, N. Grain refinement of Al–Si alloys by Nb–B inoculation. Part II: Application to commercial alloys. Materials and Design. 66 (2015) 376-383, 2015. UK, 2015.

[17] XU J, LI R, LI Q. Effect of Agglomeration on Nucleation Potency of Inoculant Particles in the Al-Nb-B Master Alloy: Modeling and Experiments[J]. Metallurgical and Materials Transactions A. 2021: 1-18. China, 2021.

[18] APELIAN, D. Aluminum Cast Alloys: Enabling Tools for Improved Performance. Wheeling: North American Die Casting Association. USA, 2009.

[19] MCQUEEN, J. H. SPIGARELLI, S. KASSNER, M. E. EVANGELISTA E. Hot Deformation and Processing of Aluminum Alloys. CRC Press. USA - NW, 2011.

[20] VERRAN, E. H. BATISTA, G. M. Análise dos efeitos dos tratamentos térmicos de solubilização e envelhecimento artificial sobre a microestrutura da liga de alumínio; Matéria R. J. Brasil, Rio de Janeiro, 2015.

[21] CAVALCANTE, F.F. Comportamento mecânico e tenacidade à fratura de ligas de alumínio: 2024 e 7075, submetida a diferentes tempos de envelhecimento. Dissertação, Engenharia dos materiais, UFRN. Brasil, RN, 2016.

[22] SHAKELFORD, J. F. Introdução a ciências dos materiais para engenheiros. PEARSON, 6ª ed., USA, 2008.

[23] BOLZONI, L. HARI BABU, N. Towards industrial Al-Nb-B master alloys for grain refining Al-Si alloys. J Materiares Technol . 2019; 8(6): 5631-5638, 2019.

[24] BOLZONI, L. BABU, N. H. Engineering the heterogeneous nuclei in Al-Si alloys for solidification control. Materials Today 5 (2016) 255-259. UK, November 2016.

Section 4

Cast Iron Technology

Chapter 6

'Vari-Morph' (VM) Cast Iron with Several Forms of Graphite: Technology, Properties, Application

Jerzy Zych, Marcin Myszka and Janusz Postuła

Abstract

Cast iron with mixed shape graphite and controlled fractions of individual shapes, known as VM cast iron ('Vari - Morph'), can become material for castings with special requirements. The name of the cast iron 'Vari Morph' (VM) was first proposed by the authors in 2018 at the World Congress of Foundries in Krakow. VM cast iron displays physical and mechanical properties, which cannot be achieved with homogeneously shaped graphite. Cast iron with (L – flake) + (V – vermicular) graphite is characterised by good thermal conductivity and better A5 (elongation) and Rm (tensile strength) (than grey cast iron. What is of particular interest is cast iron with a mixed form: (S – spheroidal) + (V – vermicular). Currently, research is being carried out to achieve cast iron with a high-quality index (QI) defined as Rm/HB. This paper presents the results of research of physical (thermal conductivity), mechanical (Rm and A5) and functional properties (thermal fatigue) of VM cast iron. The ultrasound technique was applied for assessing the graphite compactness degree (ξ): ultrasonic wave speed $C_L = f(\xi)$, damping factor $\alpha = f(\xi)$. The article also presents the correlations between the above-mentioned parameters, as well as describes the technology used to produce VM cast iron and possible areas of application of the material.

Keywords: cast iron, graphite shape, quality index, thermal fatigue, castings of special destinations

1. Introduction

Cast iron with carbon in the form of graphite, known as grey cast iron, is the most commonly used material in the casting industry around the world. Cast iron, in terms of graphite morphology, can be divided into three groups: cast iron with flake graphite (EN ISO GJL ..), cast iron with vermicular graphite (EN ISO GJV ..) and cast iron with spheroidal graphite (EN ISO GJS- ..). Currently, the production process of cast iron does not allow for using mixed forms of graphite in the same castings. Only a small percentage of a different form of graphite is acceptable (for SG cast iron—up to 5.0%, for VG cast iron—up to 20%). Production technologies for individual types of cast iron have been developed to achieve such homogeneous graphite structures.

 IntechOpen

In many cases, the introduced restrictions, and the need to maintain a highly homogeneous graphite morphology (flake, vermicular or spheroidal), considerably limit the possibility to take advantage of the numerous beneficial properties of cast iron. Graphite morphology affects all properties and characteristics of cast iron, including physical, mechanical, technological, functional and operational properties. However, such limitations have no reasonable justification. The conducted research [1–6] points to the possibility of producing cast iron with mixed graphite morphology and has confirmed the possibility of controlling this morphology in a deliberate and predictable manner. **Figure 1** presents cast iron with a mixed graphite form, which has been named Vari Morph (VM). The EN ISO 945 standard provides for only three types of graphite forms in raw castings that have not been heat-treated. The others, crossed out in **Figure 1a**, include the forms of graphite after heat treatment of cast iron. VM cast iron includes both the groups of graphite forms shown in **Figure 1b** and **c**. The idea of VM cast iron is to create a smooth transition of graphite forms from flake, through mixed: flake + vermicular (**Figure 1b**) and mixed vermicular + spheroidal, up to fully spheroidal. This smooth transition creates the possibility of taking advantage of the whole range of physical and mechanical properties. In the case of standardised cast iron grades, the changes of properties have a step-like nature.

The research described in this paper has found a significant impact of graphite morphology on the whole range of cast iron properties and has confirmed the possibility of controlling these properties. One of the most important results that can be achieved is the possibility to produce grey cast iron with relatively good strength properties (R_m - 300 ÷ 400 MPa) and, at the same time, low hardness (HB - 160 ÷ 180). In conventional cast iron with homogeneous flake graphite (EN GJS), to achieve the described strength, the pearlitic form of metal matrix is used, which increases hardness to HB - 230÷240. Contemporary methods of machine castings, with the use of high-speed CNC machines, have been adapted to materials with reduced hardness. In the case of iron castings, hardness below 200HB is required. If only flake graphite

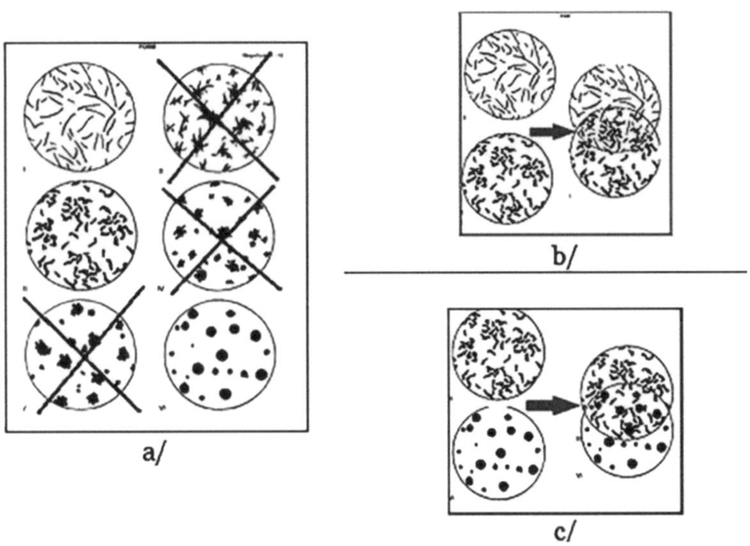

Figure 1.
Types of graphite morphology:/a – According to EN ISO 945, b/ cast iron with mixed forms of flake and vermicular graphite, c/cast iron with mixed forms of vermicular and spheroidal graphite [7, 8].

is used, it is very difficult to achieve these parameters (Rm and HB), while for higher Rm values, it is impossible. With hardness values significantly exceeding 200HB, the machining speed of CNC machines must be reduced, which is not profitable. This has been confirmed empirically by research [9] in a foundry that performs CNC machining of its own castings. Cast iron quality is often determined by the Quality Index (QI), defined as: IQ = Rm/HB. Grey cast iron has a relatively low QI value of 1.0÷1.4, while for ductile cast iron, it is approx. 2.5÷3.0. By controlling graphite morphology and allowing for mixed forms in the structure of castings, it is possible to fill the QI range between these two extremes, obtaining values in the range 1.0 ÷ 2.5. Although vermicular cast iron partly fits in between grey and spheroidal cast iron, the requirement for 80% vermicular graphite, imposed by the European standard, on the one hand, complicates the production technology, while, on the other, limits the freedom to take advantage of numerous cast iron properties, including achieving a better QI value.

Producing cast iron with other than flake graphite forms requires out-of-furnace secondary processing of liquid metal, to which small amounts of additives (inoculants, nodularisers) have been introduced. The possibility to control graphite morphology in a wide range has been confirmed by such authors as [1–6]. The mentioned research concerned the increased resistance of cast iron to thermal fatigue by controlling the wide range of graphite morphology: from flake to spheroidal [10–13]. In addition, as evidenced by the existing analytical research, cast iron with non-homogeneous graphite morphology is characterised by above-standard properties that cannot be achieved with homogeneous forms. Studies on the impact of the graphite shape factor ξ (degree of compactness) on the resistance of cast iron to thermal fatigue indicate other close correlations between other properties of cast iron and the shape factor ξ, including: $Rm = f(\xi)$, $A = f(\xi)$, $E = f(\xi)$, α (vibration damping) $= f(\xi)$, λ (thermal conductivity) $= f(\xi)$. The shape factor has been defined in numerous ways. According to one definition, it can be explained as the area occupied on a metallographic section by the graphite precipitate to its perimeter when squared. The authors of this article propose to define the graphite shape factor 'f' for VM cast iron as a weighted average of the values of the indexes for the flake, vermicular and spheroidal forms of graphite, considering the percentage share of each form in the total set of particles visible on the metallographic section. Because most cast iron properties are determined as a function of the graphite shape factor, the correct determination of this value is of key importance for the accurate description of Vari Morph (WM) cast iron characteristics.

2. Technologies of Vari Morph (WM) cast iron production

VM cast iron can be produced in several ways, similar to the production of ver micular or ductile cast iron. The technologies for producing cast iron with mixed form of graphite are similar to those for vermicular/spheroidal cast iron and include the following solutions:

1. Production of cast iron with a limited sulphur content (S < 0.01%) in the initial state, by introducing a strictly controlled amount of nodulariser (Mg < 0.03%). The introduction of the controlled quantity is achieved using three methods:

   a. Using the PE (flexible core wire) method with FeSiMg low magnesium alloy. An experimental test stand for this method will be constructed as part of the project at AGH University of Science and Technology.

b. Using low-magnesium alloys dosed at the bottom of sealed ladles (slender ladles – Tundish method).

c. Inmold technology, with placement of low magnesium alloys dedicated to Inmold technology in reaction chambers.

2. Production of cast iron using a constant amount of magnesium alloy and varying amounts of sulphur. Technology adapted to the production of cast iron in cupola-type furnaces. Testing will be carried out in Tundish-type slender ladles.

3. Cast iron production using anti-spheroidising elements, mainly titanium (Ti), introduced in strictly controlled amounts. Testing in Tundish-type ladles.

4. Production of cast iron using alloys containing additives of rare earth elements (Mischmetal), currently dedicated to the production of vermicular cast iron.

Methods most widespread in industrial practice include:

a. Controlled magnesium content method. Pure magnesium or magnesium alloy is introduced into the liquid metal in an amount less than required to obtain spheroidal graphite. In vermicular cast iron, the final Mg content varies between 0.01 and 0.03% (**Figure 2**). After the addition of magnesium, inoculation is carried out, mostly with ferrosilicon-based inoculants, similar to the production of ductile iron.

The method is quite difficult in actual industrial practice, as the range of magnesium content in cast iron, at which graphite crystallises in vermicular form, is very narrow. To produce cast iron with flake + vermicular graphite, the magnesium content must be kept below 0.02%. In turn, to produce cast iron with vermicular + spheroidal graphite, the magnesium content should be kept in the range Mg = 0.02–0.03%. The above-mentioned values apply to cast iron with low sulphur content (S < 0.01%), whereby magnesium is not used for its desulphurisation.

An excessive amount of Mg results in the formation of spheroidal graphite, while an insufficient amount Mg creates flake graphite. Such cast iron is also more prone to the fading of the vermiculisation effect compared with cerium-treated cast iron. To obtain vermicular graphite in thick cast sections, the time from magnesium

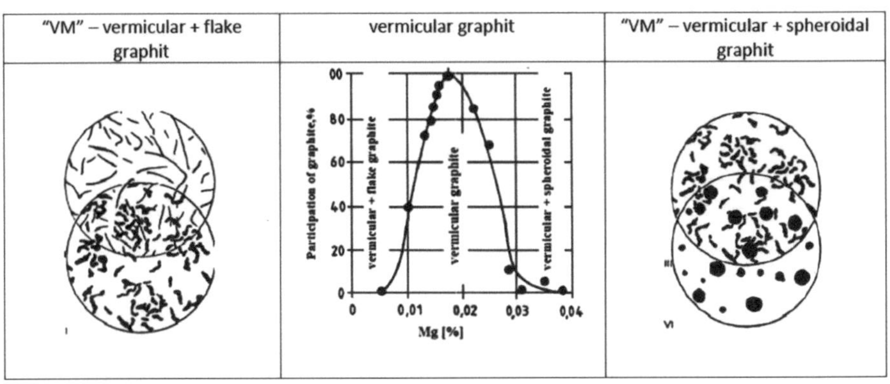

Figure 2.
Principle of obtaining vermicular graphite and Vari morph cast iron by the control magnesium content method [14].

inoculation to pouring must be kept to the minimum. In castings with different wall thickness, the method is impracticable.

b. Simultaneous introduction of magnesium and titanium. The method consists of introducing into cast iron an anti-spheroidising element, mainly titanium, with a small amount of cerium, simultaneously with magnesium. The addition of titanium extends the range of magnesium content in cast iron, at which vermicular and mixed-form graphite is formed.

c. To facilitate application of this method, a special alloy has been developed and patented: 4–5% Mg; 0.4÷0.7% MZR; 4.0 ÷ 4.5% Ca; 8.5 ÷ 10.5% Ti; 1÷ 1.5% Al.; 48÷52% Si. This alloy melts at a temperature of 1100°C. The alloy is introduced into cast iron in the same way as in the case of ductile cast iron production. Subsequently, inoculation of cast iron in the ladle with FeSi is required. The composition of the cast iron used as a starting material should be close to eutectic, and the sulphur content should not exceed 0.035%. Cast iron temperature at the time of alloy introduction should be 1450–1500°C. At this temperature and with a sulphur content of 0.035%, the amount of alloy added should be in the range of 1.5–1.7% of cast iron weight. At temperatures below 1430°C, the amount of spheroidal graphite in cast iron increases. A disadvantage of the method is the introduction of 0.1 ÷ 0.15% Ti into cast iron, which remains in the foundry circulation scrap. In addition, the machinability of castings deteriorates.

d. The introduction of Mishmetal alloy. The method consists in introducing a mixture of rare earth elements into cast iron with low-sulphur content (<0.02%S), e.g. Mishmetal alloy based on cerium (30 ÷ 50%) with the addition of: yttrium, lanthanum, praseodymium, neodymium, gadolinium and other elements. This has been developed for the production of vermicular cast iron but can also be used for the production of VM cast iron. The cast iron used as a starting material should have a chemical composition similar to that of ductile cast iron. The carbon content should not exceed 3.7%. Depending on the wall thickness of the casting, the silicon content can be up to 3.0%. Typically, the Sc eutectic saturation coefficient (the degree of eutectiveness (S_C)) of this type of cast iron is approx. 1.0 or slightly above 1.0. To obtain a ferritic structure in the cast state, pig iron with low contents of manganese (<0.3%), phosphorus (<0.05%) and sulphur (<0.02%) should be used in the smelting process. Mishmetal alloy can be added to the liquid metal stream in the form of rods. Due to the low melting point of the alloy (790÷850°C) and the high boiling point of the components (approx. 3000°C), it easily dissolves in cast iron without pyrotechnic effect and the emission of fumes. In turn, a disadvantage of this method is the significant tendency of the cast iron to form cementite (chill).

e. PQ-CGI Inmold process This method has been developed and patented by NovaCast. The PQ-CGI Inmold process is based on precise control of the oxygen content (in cast iron) and the formation of crystallisation nuclei. The process uses the PQ – CGI system, which is based on a modern thermal analysis method.

f. Sinter – Cast technology: the method takes advantage of the effects of vermiculisation (with controlled quantity of magnesium) and inoculation treatment. The use of two flexible wires for the procedure in industrial conditions, one of which contains a magnesium core and the other an inoculation core, ensures the efficiency and repeatability of this technology.

Master alloy	Si	Mg	Ce	Ca	Al
COMPACTMAG	44–48%	5–6%	5–7%	0.8–1.2%	1.0% max

Table 1.
Chemical composition of the COPACTMAG master alloy [15].

g. Elkem method (COMPACTMAG alloy). The method uses a new alloy developed by the Elkem company to produce vermicular cast iron. Its composition is listed in **Table 1**. The alloy can also be used in the production of VM cast iron with mixed-form graphite. This alloy has a magnesium content of 5÷6% and an addition of cerium (5–7%). It is used for vermiculisation in the treatment ladle, using the Sandwich method and its variations.

2.1 Proprietary technologies to produce Vari morph cast iron

A method for producing VM cast iron was selected based on preliminary tests. The tests were initially conducted in the experimental foundry of the Faculty of Foundry Engineering of the AGH University of Science and Technology in Kraków and then in a selected industrial foundry. The controlled Mg content method was selected, and the nodularisation process was carried out using a ladle with a tight lid (Tundish technology) and a flexible core wire (PE) method. Schematic illustration of the process is presented in **Figure 3**. In industrial conditions, the nodularisation process was also carried out using the Inmold method. Using this technology, cast iron with mixed-form graphite (Vari Morph) is obtained by introducing a lower amount of Mg

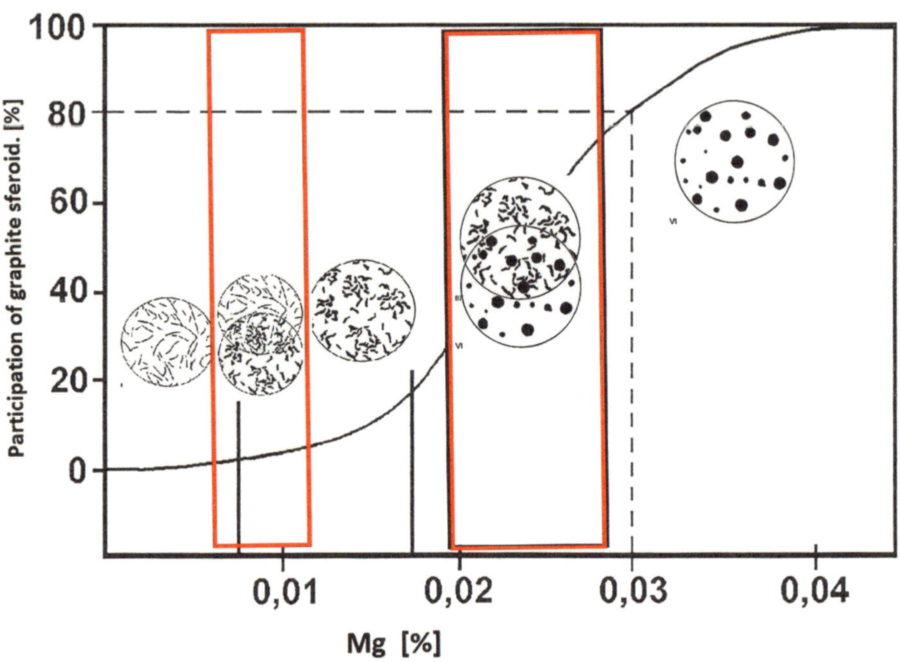

Figure 3.
Schematic diagram of the Vari morph cast iron production method with the introduction of strictly controlled amounts of Mg.

compared with the amount used in ductile iron production. Therefore, the secondary (out-of-furnace) treatment itself, consisting of the introduction of strictly controlled amount of Mg, can be treated as cast iron nodularisation process. This term is used to describe it in this paper. In the case of VM cast iron, nodularisation is incomplete, and the graphite morphology is variable, covering three basic forms in which it occurs in cast iron: flake (L), vermicular (V) and spheroidal (S).

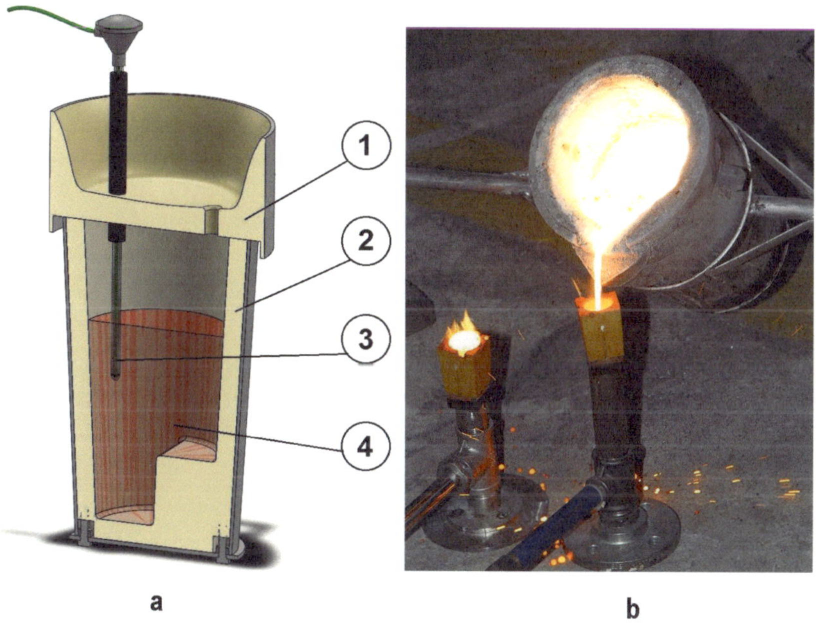

Figure 4.
Tightly covered ladle (Tundish): a/ drawing: 1 – Cover, 2 – Ladle, 3 – Thermocouple, 4 – Liquid metal; b/ pouring samplers from the ladle for TDA analysis.

Figure 5.
Experimental stand for cast iron nodularisation with the PE method): a/ drawing; b/ processes performed by cored wire method – Precision dispensing Mg.

Figures 4 and **5** show the test stands in the experimental foundry, while **Figures 6** and **7** show the industrial version. **Figure 4** shows a drawing of the experimental Tundish ladle and the pouring of samplers for thermal derivation analysis (TDA).

In the second technological solution for the production of Vari Morph cast iron, nodularisation of cast iron was carried out by the PE (flexible core wire) method.

The experimental test stand is shown in **Figure 5**. The cast iron nodularisation process involves the introduction of a steel wire (thin-walled tube) filled with FeSiMg alloy containing magnesium, usually in the amount of 16–23%, into the liquid metal. The process was carried out with full control over the rate of rod immersion in the metal and the amount of alloy, or more precisely Mg, introduced into cast iron. Production of VM cast iron requires strict control of the amount of Mg introduced into cast iron.

Experimental production of VM cast iron has also been carried out under industrial conditions as part of NCBiR projects [9]. Test ingots and castings were made in moulds with a vertical parting line, on the DISA MATIC line (**Figure 6**), and moulds with horizontal parting line on the FBO automatic moulding line (**Figure 7**).

The purpose of the tests conducted in the experimental foundry as well as under industrial conditions was the assessment of the possibility to produce VM cast iron

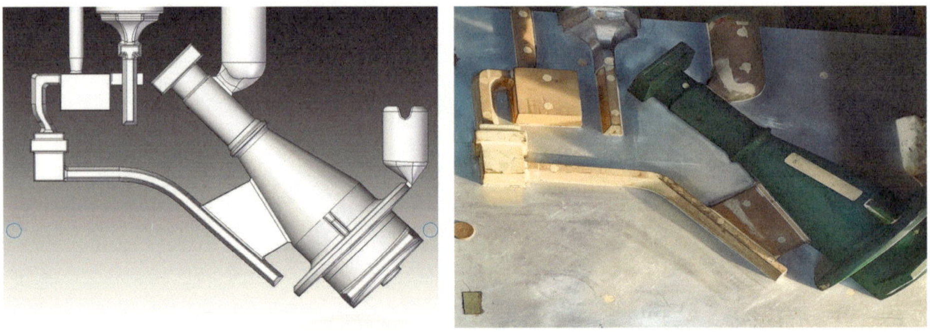

a b

Figure 6.
Experimental technology of VM cast iron production using the Inmold process on a DISA MATIC line: a/ general concept of the technology, b/ view of model plate [(according to patent [16]).

a b

Figure 7.
Model plate and mould for research on the production of VM cast iron using the Inmold technology on the FBO line; a/ upper model plate, b/ lower mould [9].

in a manner ensuring repeatability and the required structure of graphite in the casting, as well as the assessment of the possibility to take advantage of selected physical, mechanical and functional features and properties of cast iron. Comparison of the three methods of cast iron nodularisation allowed us to verify them in terms of efficiency of the process of magnesium introduction from FeSiMg alloys into cast iron (degree of magnesium assimilation). The highest degree of Mg assimilation is achieved with Inmold technology (65–75%), followed by Tundish ladle nodularisation technology (55–60%), while the lowest was with the PE flexible core wire technology (30–35%). The Mg assimilation degree for each technology depends on several factors, including the temperature of the cast iron, sulphur (S) and oxygen (O2) content. Nevertheless, the following estimates can be accepted as a guideline.

The authors of this paper compiled the results of research on plain cast iron with a chemical composition close to eutectic composition (Sc ~1.0; CE ~ 4.30) with reduced manganese content (Mn < 0.25). In most cases, cast iron had a ferritic matrix, which allowed us to assess the impact on physical and mechanical properties of graphite morphology rather than of the type of matrix.

The results allowed determining several empirical correlations between the graphite shape factor and selected properties.

A group of cast iron grades, the chemical composition of which was close to the eutectic value, was subjected to analysis (C = 3.3÷3.6%, Si =2.6÷2.95%, Mn < 0.25; P < 0.02%; S < 0.01%; Mg =0.005÷0.040%).

2.2 Determination of the graphite shape factor

The graphite shape factor, a key parameter describing the structure of Vari Morph cast iron with mixed-form graphite, requires more detailed analysis. The literature on the subject [1–5] proposes various factors describing numerically the shape of a single precipitate. Some authors [10, 11] used the ξ factor (formula 1). Regarding VM cast iron, the arithmetic or weighted average of this actor was used after counting the percentage share of individual graphite forms.

The graphite shape classification was based on the indicator shape of graphite ξ, which was defined by formula 1 [17]:

$$\xi = \frac{A_i}{P_i^2} \tag{1}$$

where:

A_i, P_1 – surface (m^2) and diameter (m), respectively, of a single precipitate,
i – number of graphite precipitates. Indicator ξ takes the following values:

0÷0.03 (ξ < 0.03) for flake graphite, 0.035÷0.065 for vermicular graphite, 0.065÷0.08 (ξ > 0.065) for nodular graphite.

The computer-aided methodology of determining the shape indicator was developed, with using the Image J program. This is a new method especially suitable in quantitative determinations of this indicator of cast iron grades having different graphite forms. This method is based on the concept of calculating the mean weighted value of the shape indicator, and all visible separations on the microstructure are included in the calculation, regardless of their size and shape.

Microscopic images were subjected to the stereological analysis by means of the program Image J for pictures analysis. The microstructure image in digital recording and the image processed by the Image J program are shown in **Figure 8**, as an example.

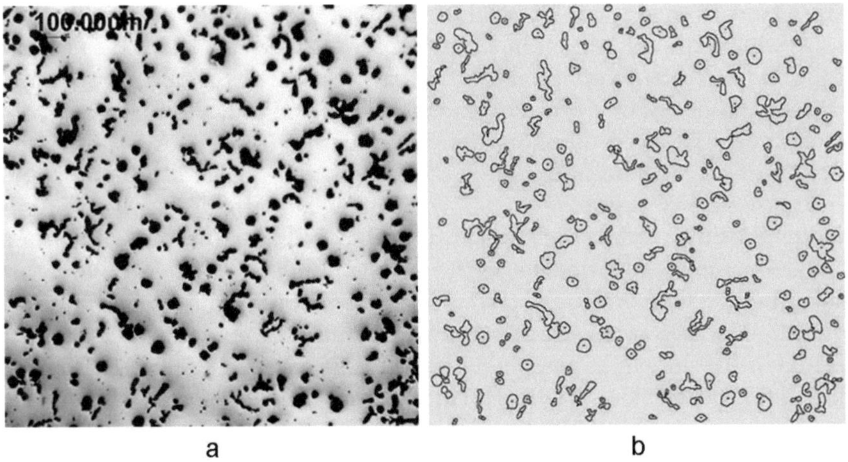

Figure 8.
Microstructures of cast iron (VM) – With a mixed graphite form. a/ microstructure image. b/ image processed by the image J.

Each graphite precipitate is recorded by the program by means of the graphite shape indicator included in Eq. (2) [5]. Graphite single precipitation shape index 'f' is defined as the ratio of the graphite precipitation area to the circle area with a diameter equal to the largest particle size (graphite) (**Table 2**).

$$f = \frac{A_V}{A_c} \tag{2}$$

The assessment of the shape index in VM cast iron, in which different forms of graphite occur next to each other, is always simplified and incomplete, regardless of the calculation methodology. Before making the selection, the authors made a number of comparisons, trying to correlate the value of the determined shape index with the selected physical and mechanical properties of cast iron. Better correlation of the results was obtained with the use of the 'f' index described by Eq. (2). The shape indices are determined for a flat 2D structure image, and the cast iron properties are related to the three-dimensional image of the 3D graphite particles. At the present stage of the description of graphite particles, such methodology is commonly used. Eq. (2) describes, on the one hand, the level of separation compactness in a simpler way, and on the other hand, it accentuates the scale of the distance between the shape and the spherical shape. As indicated in **Table 2**, the value of this indicator (f) changes nearly eight times when switching from the flaky to the ball form. In the case of the shape indicator (...), with a similar change in the shape of the graphite particles, the indicator value changes only threefold. The conducted analysis and the comparisons made prompted the authors to choose the 'f' index to describe the sought dependencies for VM cast iron.

The procedure of determining the mean shape indicator was performed in subsequent steps in which the written below values were determined.

- Number of graphite precipitates per mm^2 (with omitting precipitates being on the picture edges and the ones which surface is smaller than 0.0785 µm), **Figure 9**;

DOI: http://dx.doi.org/10.5772/intechopen.102045

Graphite section	Calculation formula					
$\frac{\textit{Graphite surface area}}{\textit{Circle surface area}} \times \%$	$f = \frac{A_{graphite}}{A_{circle}} \times 100$	90.8	79.5	57.5	34.6	12.7
$\frac{\textit{Graphite surface area}}{\textit{Circle surface area}}$	$f = \frac{A_{graphite}}{A_{circle}}$	0.91	0.80	0.58	0.35	0.13
$\frac{\textit{Graphite surface area}}{\textit{Graphit diameter}(P)^2}$	$\xi = \frac{F_{graphite}}{P^2}$	0.080	0.070	0.060	0.045	0.030

Table 2.
Graphite shape index values ξ, i 'f' calculated according to various equations.

Figure 9.
Scheme of shapes of graphite precipitates cross sections [10].

- Surface percentage fraction of graphite;
- Surface of each precipitate;
- Diameter of the circle on the precipitate corresponding to its highest dimension.

The graphite shape indicator f was determined for each precipitate, acc. to [12], in accordance with Eq. (2), where:

A_v — surface of the graphite precipitate [mm^2],

A_c — area of the circle of a diameter equals the highest particle dimension [mm^2].

The shape indicator 'f' theoretically changes from 0.0 to 1.0, while under real conditions from app. 0.20 to 0.95.

It was assumed that the shape indicator of graphite is changing within ranges:

Graphite shape	Number of precipitates	% Fraction of the range	Mean 'f' of the range	Indicator 'f'
Flake	31	12.0	0.27	0.62
Vermicular	106	41.1	0.49	
Spheroidal	121	46.9	0.81	

Table 3.
Analysis of the shape indicator.

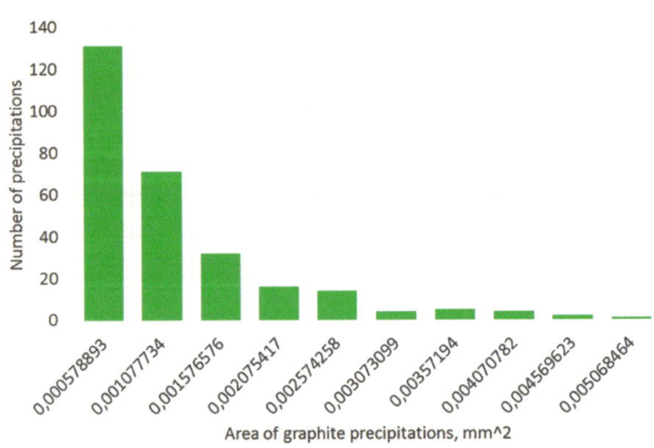

Figure 10.
Number of graphite precipitates in individual surface classes.

- 0.00–0.34 for flake graphite,
- 0.35–0.64 for vermicular graphite,
- 0.65–1.00 for spheroidal graphite.

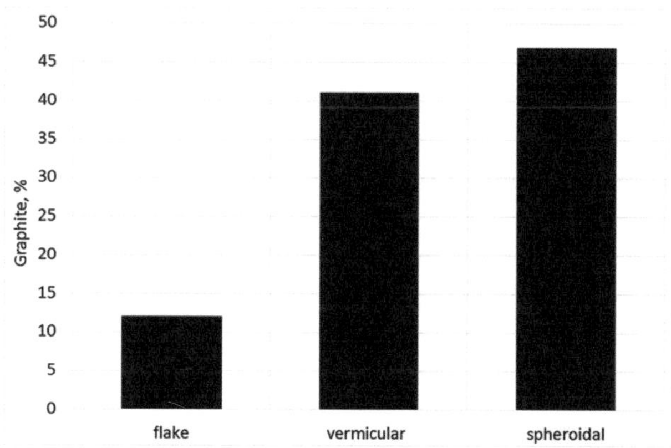

Figure 11.
Fractions of individual graphite kinds.

	The appearance of the structure and the value of the indicator 'f'		Graphite form type
1	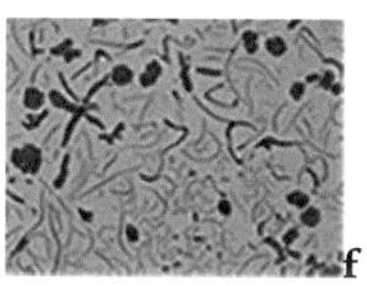 f = 0.28	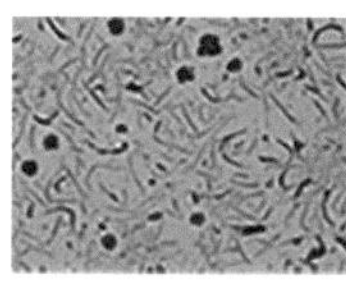 0.30	L + V + S
2	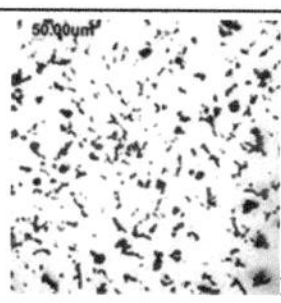 f = 0.51	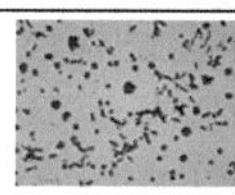 f = 0.52	L + V
3	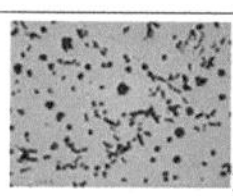 f = 0.63	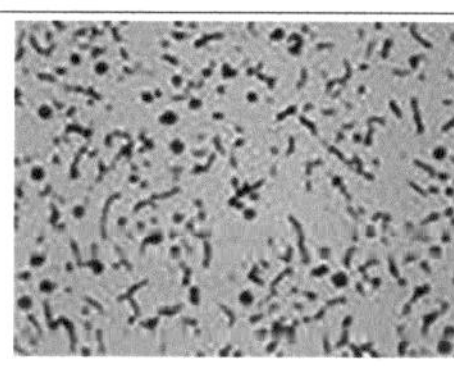 f = 0.66	V + S
4	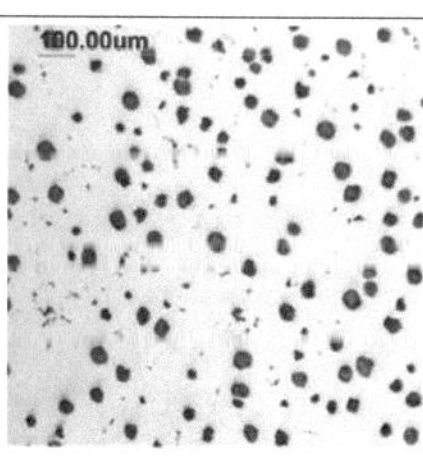 f = 0.78	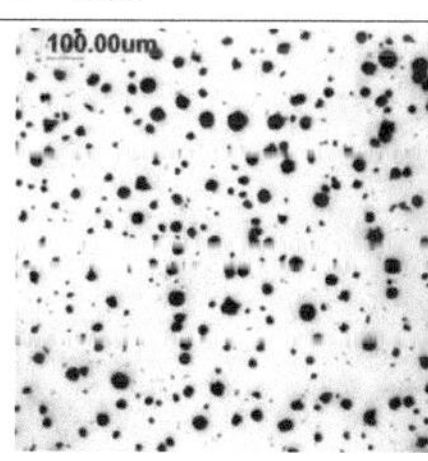 f = 0.84	S

Table 4.
Typical metallographic structures and calculated 'f' values.

In the next step, the number of precipitates of the shape indicator 'f' being in individual ranges was calculated, and their percentage fraction in relation to all precipitates was determined. For each range, the separate mean indicator f was determined. They constituted the bases for calculating the main shape indicator f, based on the weighted mean. Individual results of graphite shape analyses obtained in the assumed calculation procedure are listed in **Table 3**.

It was indicated on the basis of the performed calculations that the mean indicator of the graphite shape f = 0.62, which classifies precipitates as vermicular graphite. Simultaneously the spheroidal degree is N = 46.9%. Fractions of individual kinds of graphite shapes are shown in **Figure 10**. One of the results of the complex assessment of graphite precipitates is placing them within ranges of the shape indicator 'f', shown in **Figure 11**. In the analysed case of VM cast iron precipitates, shape indicator 'f' of which is within the range 0.22 ÷ 0.98, corresponding to the top range of the vermicular graphite, are dominating (**Table 4**). The shape indicator for each cast iron from successive melts was determined in the same way.

3. Results of own investigations

3.1 Tests of physical and performance properties

Applying the described melting methodology, the material for tests was prepared in forms of test coupons Y type, from which samples for individual tests were made. The following physical properties were assessed: specific density, ability to propagate longitudinal waves and thermal conductivity of VM cast iron. These properties of 'Vari Morph' (VM) cast iron depend on the graphite form 'f', which allows to determine dependencies: $\rho = f(f)$; $C_L = f(f)$ and $\lambda = f(f)$. The limited Mn content in cast iron allowed obtaining purely ferrite cast iron or ferrite with a small (to app. 15%) pearlite fraction. In this way, the influence of the metallic matrix kind on the tested properties of cast iron was eliminated.

The ultrasound wave speed in cast iron depends on the graphite form and changes within a wide range [16–21]. This is confirmed by the results of investigations presented in **Figure 12**. In investigations we are striving to the development of the empirical dependence $f = (C_L)$ for the whole variability range of the shape indicator. If such dependence is characterised by a high correlation coefficient (high certainty), the ultrasound technique will be used for the non-destructive assessment of the graphite shape indicator in ready castings. The results of tests, performed on a relatively not numerous group, indicate a good correlation between these two values.

Specific density (ρ) of VM cast iron was determined by the immersion method with using a strain gauge adapted to these measurements. The procedure of the sample density measuring was the standard one.

The results of investigations of the influence of the graphite precipitates shape on the density of VM cast iron are presented in **Figure 13**. It was found in metallographic tests that the graphite amount, measured as the surface taken on the polished section, was similar in all tests. This was due to maintaining the stabilised chemical composition of cast iron. It can be noticed that the transfer from flake via compact (vermicular) to the spheroidal form of graphite causes the density increase by approximately 0.15 g/cm3. The so-called compactness of material increases, which influences its certain properties such as: strength, tightness, plasticity.

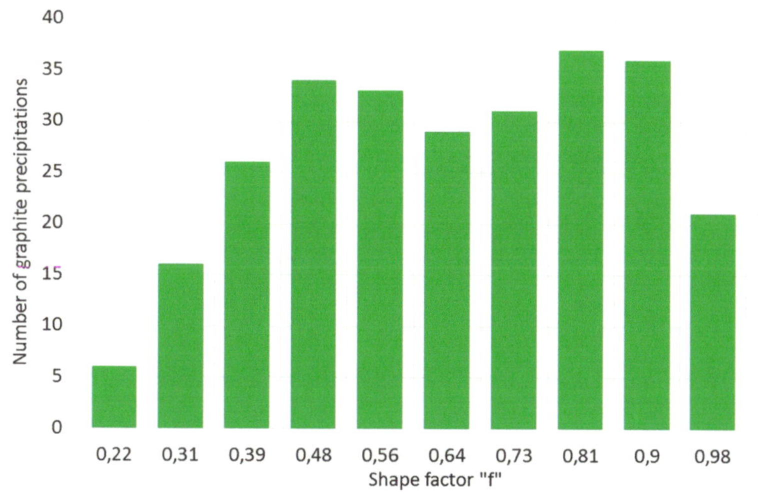

Figure 12.
Number of graphite precipitates as a function of the shape indicator.

In the future, VM cast iron is supposed to be the material, structure of which—related to graphite precipitates shapes—will be 'adjusted' to the destination of the product and the character of its operation under real conditions. One of the features, which will be controlled, is the thermal conductivity of cast iron. Investigations of the samples group obtained in successive melts were carried out on the prototyped research set-up, shown in **Figure 10**. The thermal conductivity was determined under conditions of a stable heat flow in cylindrical samples of diameter ø20mm and length 80 mm. The conductivity was determined within the temperature range T = 25 ÷ 500°C. The original research site is shown in **Figure 14**, and the obtained results are presented in **Figure 15**.

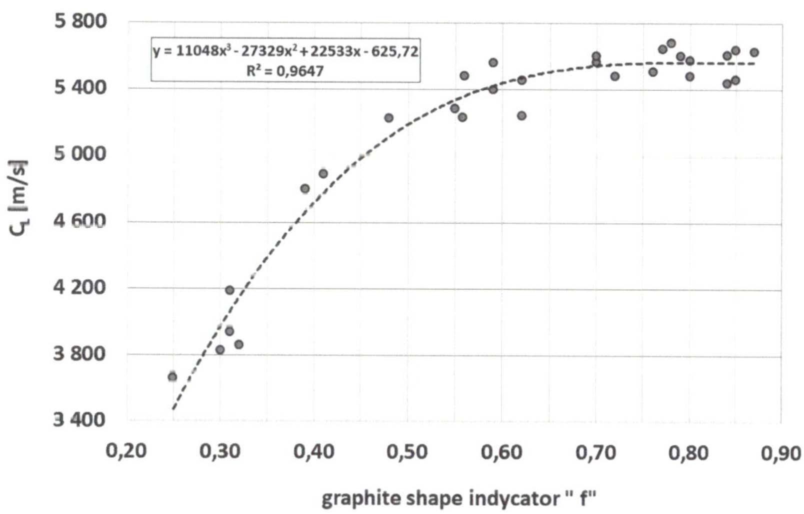

Figure 13.
Dependence of the mean shape indicator 'f' and the ultrasound wave speed C_L.

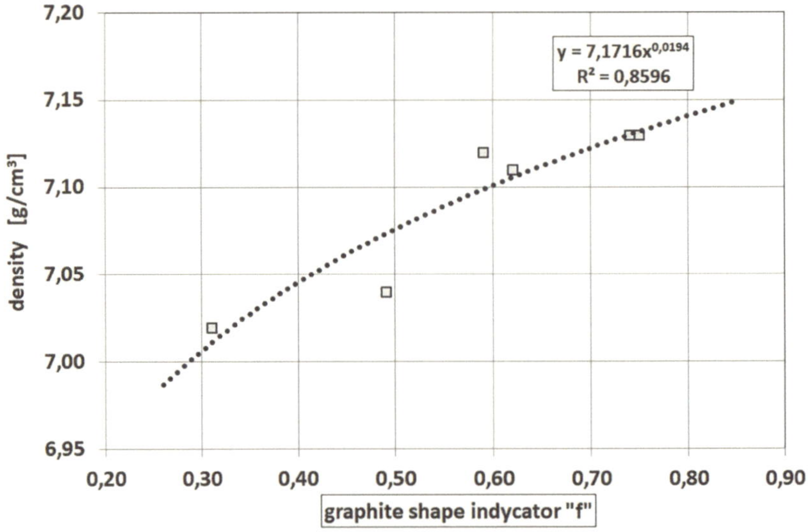

Figure 14.
Influence of the graphite form (indicator 'f') on the VM cast iron density.

Figure 15.
Research set-up for testing the thermal conductivity of materials, including foundry metals and alloys; a/view of the stand, b/preparation of the sample for testing.

It is generally known that the change of the graphite form from flake (f ~ 0.25) to spheroidal (f > 0.75) leads to decreasing of the thermal conductivity. The quantitative dependence, in which the graphite shape is described by the mean value of indicator 'f', is determined in the hereby paper.

The functional property of cast iron, which depends on the graphite form, is the heat fatigue resistance. The heat fatigue resistance tests were performed by means of the L.F. Coffin method [10, 18, 22], in which cylindrical samples are cyclically heated by resistance method. The results obtained for VM cast iron are presented in **Figure 16**. Increasing the graphite shape indicator 'f' value, increasing the compactness degree of graphite precipitates, favours increasing the heat fatigue resistance. However, it should be noticed that under real operational conditions of cyclically heated structure elements, the structures made of cast iron with more compact graphite (higher 'f' indicator) will be heated to higher temperatures, due to their low conductivity (**Figure 17**).

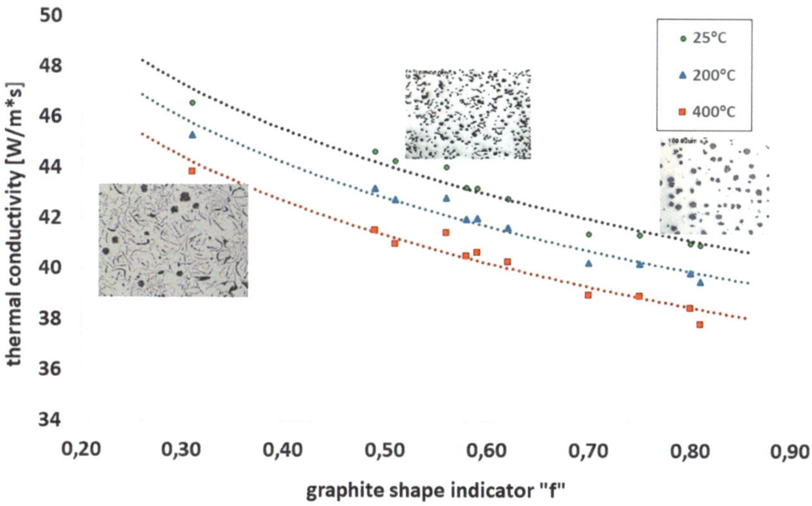

Figure 16.
Dependence between the graphite shape indicator and thermal conductivity of ferrite cast iron.

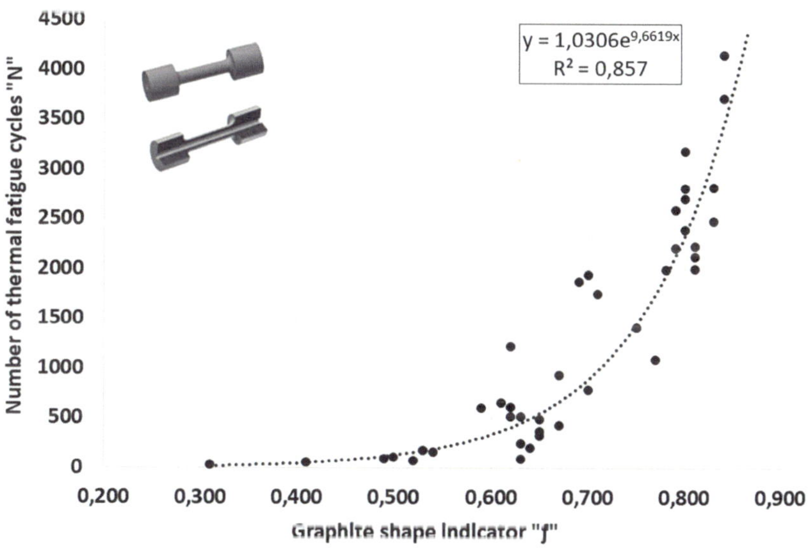

Figure 17.
Influence of the graphite shape indicator on thermal fatigue resistance of VM cast iron.

3.2 Tests of mechanical properties

The results of tests of mechanical properties on samples cut from the test ingots cast in each sample are listed in **Table 5**. Rm and A5 were determined in tensile testing, while Brinell hardness was measured on the heads of the specimens after they were broken using a 10 mm ball. The quality index (QI = Rm/HB) was then calculated based on the determined values. All the values are presented in **Table 5**. Quite interesting results, especially in terms of the QI, were obtained for a group of seven casts in which the graphite form was intentionally modified using secondary out-of-furnace

No.	f	R_m	A_5	HB	IQ
1	**0.31**	117.6	1.3	173	**0.680**
2	**0.62**	430.0	9.3	183	**2.350**
3	**0.64**	341.8	13.5	141	**2.424**
4	**0.66**	372.9	12.5	149	**2.503**
5	**0.70**	522.1	13.8	181	**2.885**
6	**0.43**	363.1	6.1	183	**1.984**
7	**0.85**	609.8	13.1	200	**3.049**

Table 5.
Properties of VM cast iron with a dominant ferritic matrix.

treatment. QI is defined as the ratio of tensile strength Rm to hardness HB. By leaving the ferritic matrix (HB < 185) and increasing cast iron strength by way of 'compacting', high values of the QI index were obtained (significantly higher than in the case of cast iron with flake graphite form).

Some of the most important mechanical properties, which change depending on the value of the graphite shape factor, are tensile strength (R_m) and plasticity (A5). When analysing the correlation between the graphite shape factor and R_m and A5 values, the impact of the metallic matrix should be eliminated. Thus, tests should be performed on samples with the same type of matrix. The result obtained for ferrite cast iron is presented in **Figure 13**. Although investigations are at their initial stage, it is already possible to determine the correlation between the shape factor 'f' and the strength Rm of VM cast iron. By changing the graphite form within the range 0.30 ÷ 0.80, it is possible to increase cast iron strength from approximately 100 MPa to more than 500 MPa.

Similarly, the strong influence of the graphite shape factor is observed at tests of cast iron elongations and A5 determining. The results are shown in **Figure 18**. Changes of the shape factor 'f' within the range 0.30 ÷ 0.80 cause the elongation

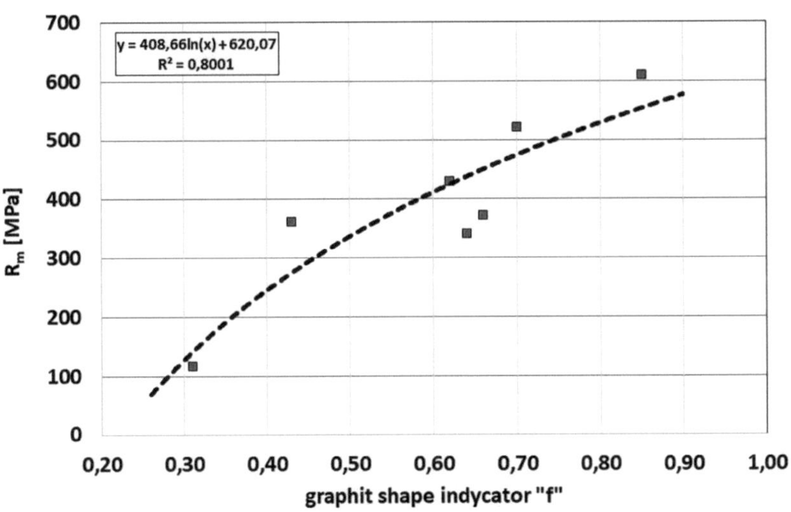

Figure 18.
Correlation between the graphite shape factor and tensile strength R_m of ferrite matrix VM cast iron.

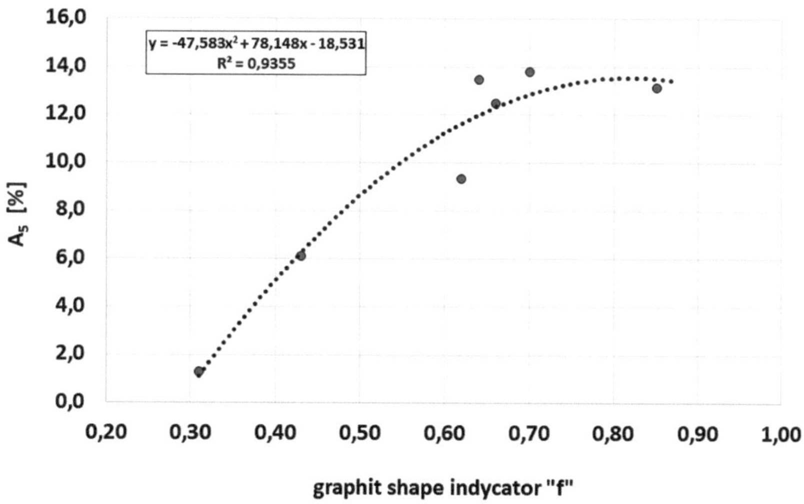

Figure 19.
Correlation between the graphite shape factor and plasticity A5 of ferrite matrix VM cast iron.

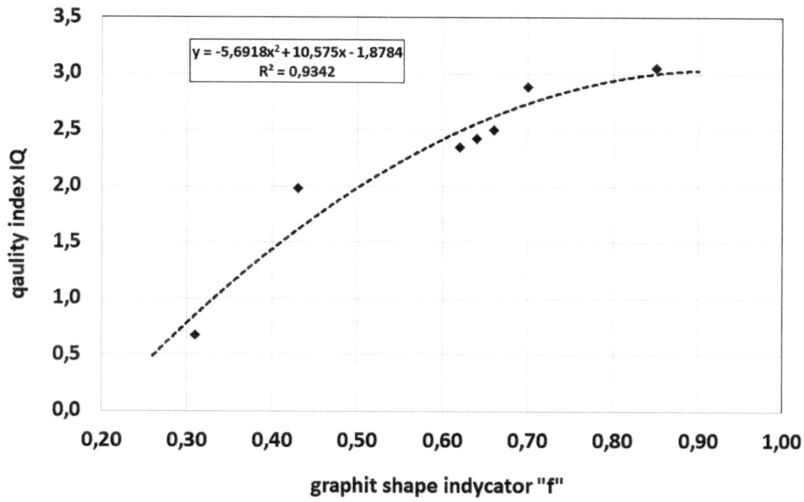

Figure 20.
Correlation between graphite shape factor 'f' and cast-iron quality index QI.

increase of VM cast iron from approx. 1.5% to approx. 14%. The direction of changes is known; thus, the tests involve the determination of empirical correlations in the form of mathematical dependencies.

The QI index is a measure of the quality of grey cast iron with mixed-form graphite, including the VM cast iron in question. The value of the quality index QI increases proportionally to the increasing value of the graphite shape factor 'f.' The empirical correlation between these values, describing the characteristics of the Vari Morph cast iron, is shown in **Figure 19**. The definition of the QI index suggests that an increase in its value means a greater increase in strength than in hardness, which is caused by changes in the metallographic structure (**Figure 20**). This trend of increasing Rm, while keeping HB constant, is particularly advantageous in terms of cast iron machining. Ferritic matrix castings, despite their high strength achieved by increasing the 'f' factor (compacting graphite precipitates), can be very easily machined with the use of CNC machining devices.

4. Examples of cast iron castings with mixed-form (Vari Morph) graphite

In industrial practice, the presence of mixed-form graphite in castings is a very common phenomenon, especially in the production of vermicular or spheroidal cast iron. Differences can be observed within a single casting, in its walls solidifying at different rates, or between castings that use cast iron with different physical and chemical properties. Naturally, the variation in the structure and form of graphite applies to those cast iron grades that have been produced using out-of-furnace nodularisation treatment in their liquid state.

However, in the authors' opinion, the mentioned cases should not be classified as Vari Morph cast iron. VM is a cast iron type with mixed-form graphite, the structure of which has been intentionally varied using the technological process of nodularisation. The characteristics of cast iron obtained by creating non-standard cast iron grades are attractive from the point of view of their practical application. Thanks to the new structure of graphite precipitates, it is possible to obtain such desirable properties of cast iron as: increased resistance to temperature changes and thermal shock, increased tightness of cast iron, increased quality index QI and better machinability with comparable or higher strength R_m.

Several examples of castings that are/should be produced from Vari Morph cast iron are presented below. **Figure 21** shows a heating furnace element operating under cyclic heating and cooling conditions. The component is subject to wear due to thermal fatigue.

Other examples of castings subject to wear as a result of thermal shock include cinder pot, steel ingot moulds (**Figure 22**) and moulds for pouring AL alloys into ingot moulds (**Figure 23**). In existing practice, these castings have been made of both spheroidal and vermicular cast iron, and metallurgical ingot moulds have also been made of grey cast iron. The spheroidal form of graphite hinders heat dissipation leading, on the one hand, to a slower casting process of the ingots, and, on the other, to greater overheating of the moulds and cinder pots, which is also an undesirable feature that accelerates their wear. In all these cases, Vari Morph cast iron with mixed-form graphite—vermicular + spheroidal in similar proportions—would be a better material. Such cast iron should be characterised by the graphite shape factor 'f' in the range of f = 0.65–0.75.

Figure 21.
Heating furnace element: Casting made of MV cast iron operating under conditions of cycling heating.

a b

Figure 22.
Metallurgical equipment (a/ cinder pot, b/ casting mould) made of ductile or vermicular cast iron operating under cyclic heat shock conditions.

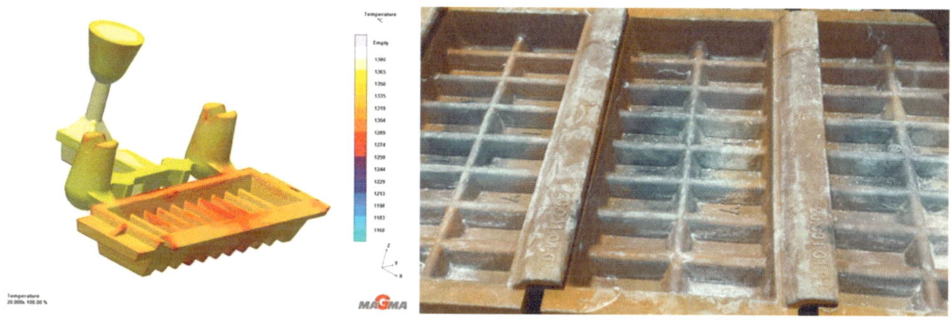

a b

Figure 23.
Mould for pouring AL alloys and casting ingots (a/ casting technology, b/ working surface). Material for casting moulds—Ductile cast iron or, more rarely, vermicular cast iron.

5. Conclusions

The conducted research has confirmed the technological possibility of producing Vari Morph cast iron with mixed-form graphite in a controlled manner. It can be produced in several ways, both in experimental foundries and under industrial conditions. VM cast iron has several advantageous properties, including a high QI. It allows us to fill the gaps between the existing standardised cast iron grades with homogeneous graphite morphology: flake, vermicular or spheroidal. VM cast iron can be a great material for structures and castings that require above-standard properties that cannot be achieved with homogeneous graphite, such as application for thermal shock operating conditions, increased tightness of castings, etc. In the authors' opinion, VM cast iron is a promising material of the future, filling a gap in the offer of the existing cast iron grades.

All properties of cast iron with a homogeneous ferritic matrix have been described in this paper as a function of the graphite form factor, the determination procedure of which has also been developed by the authors and verified by means of the described tests.

The research conducted has allowed us to determine empirical correlations between the graphite shape factor and the selected properties and functional features of VM cast iron.

These correlations involve:

- physical properties (specific density, thermal conductivity, ultrasound wave speed): $\lambda = f(f)$; $CL = f(f)$; $\rho = f(f)$,

- mechanical properties (strength (Rm; A5)): $Rm = f(f)$; $A5 = f(f)$, $IQ = f(f)$,

- functional properties (thermal fatigue resistance) $N = f(f)$

- At the current stage of work on the Vari Morph cast iron production technology, it can be stated that in the case of medium-size castings, the easiest way to process the process is to use a tightly covered Tundish ladle and a controlled amount of added Mg

- In the case of small castings produced on machine moulding lines, the Inmold technology is a favourable solution for the production of VM cast iron, especially in the case of serial production of castings with the use of spheroidising automotive dedicated to this technology.

Acknowledgements

This research was conducted within the project of: POIR 01.01.-00-0042/17.

Author details

Jerzy Zych[1]*, Marcin Myszka[1] and Janusz Postuła[2]

1 Faculty of Foundry Engineering, AGH University of Science and Technology, Krakow, Poland

2 FANSULD Cast Iron Foundry, Końskie, Poland

*Address all correspondence to: jzych@agh.edu.pl

References

[1] Jura S, Jura Z. Influence of chemical constitution and graphite spheroidization grade on mechanical properties of cast iron. Archives of Foundry Engineering. 2001;**1**(2/2):167-174

[2] Jura S, Jura Z. Influence of stereological functional parameters of graphite on ductile cast iron's mechanical properties. Archives of Foundry Engineering. 2001;**1**(2/2):175-185

[3] Stawarz M, Jura S. Stereological parameters of graphite and chemical constitution determining ductile cast iron's mechanical properties. Archives of Foundry Engineering. 2002;**2**(4):446-453

[4] Zych J. Postuła J. Cast Iron "Vari-Morph" (VM) with Graphite of Several Forms – Material for Castings of Special Destinations "Creative Foundry". 23-27 September 2018; 73rd World Foundry Congress. Kraków, Poland: congress proceedings [Cz. 2], Scientific and technical / Polish Foundrymen's Association, World Foundry Organization. — [Kraków] : Stowarzyszenie Techniczne Odlewników Polskich, 2018. pp. 149-150

[5] Patiero V, Tosto S, Natale E. Microstructural characterization and mechanical properties of various grade of nodularity in S. G. iron. In: 44 I. F. C. (International Foundry Congress), Paper No 4, Florence. 1977

[6] Zych J, Myszka M, Kaźnica N. Control of selected properties of "Vari-Morph" (VM) cast iron by means of the graphite form influence, described by the mean shape indicator. Archives of Foundry Engineering. 2019;**19**(3):43-48

[7] ISO/TR 945-2, Microstructure of cast irons — Part 2: Graphite classification by image analysis

[8] ISO 945-1, Microstructure of cast irons — Part 1: Graphite classification by visual analysis

[9] Project: POIR 01.01.-00-0042/17. Development of a technology for the production of cast iron with controlled graphite morphology and its application for the production of high-performance castings above-standard technological, functional and operational properties. Available from: https://www.gov.pl/web/ncbr-en

[10] Zych J. The role of graphite morphology in forming of thermal fatigue resistance of cast iron. Inżynieria Materiałowa. 2015;**36**(5): 343-347

[11] Zych J. Cast Iron Thermal Fatigue Resistance – Influence of Graphite Shape. Jakośći innowacyjność w procesach odlewniczych, obróbce cieplnej i plastycznej oraz w wytwarzaniu materiałów kompozytowych : praca zbiorowa / pod red. Łukasza Miki ; Politechnika Krakowska im. Tadeusza Kościuszki. Kraków: Wydawnictwo PK, 2015:51-64

[12] Francis Walton C, Opar TJ. Iron Castings Handbook. Covering data on Gray, Malleable, Ductile, White, Alloy and Compacted Graphite Irons. Iron Castings Society. 1981:831

[13] Postuła J. Thesis-Department of Foundry. Kraków-Poland: AGH; 2017

[14] Nachtelberger E, Puhr H, et al. Stand der Entwicklung von Gusseisen mit Vermiculagrahit–Herstellung, Eingenschaften und Anwendung. 22. Berlin: Giesserei-Praxis; 1982. pp. 359-372

[15] Available from: http://metals-minerals.pl/odlewnictwo-zeliwa/materialy-metalurgicze-elkem/elkem-zaprawy-magnezowe/. 2022

[16] Zych J, Żyrek A. Patent PL (11) 230233 (21); (13) B1; Application Nr: 414413 (22), 19.10.2015

[17] Sękowski K, Machynia E. Prace Inst. Odlewn. 1970;**20**(2):149

[18] Coffin LF Jr. A study of the effects of cyclic thermal stress on a ductile metal. Transactions of the American Society of Mechanical Engineers ASME. 1954;**76**:931-950

[19] Krautkramer J, Krautkramer H. Ultrasonic Testing of Materials. New York: Springer-Verlag; 1983

[20] Orłowicz AW. Zastosowane ultradźwięków w odlewnictwie. (Applied Ultrasound in Foundry) Monografia. Katowice: PAN; 2000

[21] Zych J, Falęcki Z. Testing of grey cast iron mechanical properties after the ultrasonic non-destructive method. Zeszyty Naukowe AGH, Metalurgia i Odlewnictwo. 1991;**18**(2):221-233

[22] Zych J. The impact of some selected factors on cast iron thermal fatigue. Science Notebooks - AGH, Metal. i Odlew. 1989;**15**(21):131-145